CONTENTS

vol. 5

BASIC RADIO
vol. 5
REVISED SECOND EDITION

MARVIN TEPPER

HAYDEN BOOK COMPANY, INC.
Rochelle Park, New Jersey

ISBN 0-8104-5925-6
Library of Congress Catalog Card Number 73-6498

Hayden Book Company, Inc.
50 Essex Street, Rochelle Park, New Jersey 07662

Printed in the United States of America

1 2 3 4 5 6 7 8 9   PRINTING

74 75 76 77 78   YEAR

PREFACE

Basic Radio is a course in communications electronics, as distinct from a general course in electronics. The text deals with the circuitry and techniques used for the transmission and reception of intelligence via radio energy. Assuming no prior knowledge of electricity or electronics, the six volumes of this course "begin at the beginning" and carry the reader in logical steps through the study of electricity and electronics required for a clear understanding of radio receivers and transmitters. Illustrations are used profusely to reinforce the highlights of the text. All examples given are based on actual or typical circuitry to make the course as practical and realistic as possible. Most important, the text provides a solid foundation upon which the reader can build his further, more advanced knowledge.

No prior knowledge of electricity or electronics is required for the understanding of this series. Because this series embraces a vast amount of information, it cannot be read like a novel, skimming through for the high points. Each topic contains a carefully selected thought or group of thoughts, so that each unit can be studied as a separate subject. Mathematics is kept to a minimum and, where necessary, the mathematical methods are fully explained.

This Revised Edition of *Basic Radio* retains the structure of the First Edition. Volume 1 treats d-c electricity. The slightly more involved subject of a-c electricity is presented in Volume 2. Equipped with this information, the reader is ready to study the operation of electron tubes and electron tube circuits in Volume 3, including power supplies, amplifiers, oscillators, etc. The components of electronic circuitry presented in Volumes 1 through 3 are assembled in Volume 4, which discusses the complete radio receiver, AM and FM. Volume 5 gives special attention to the theory and circuitry of transistors and integrated circuits. Volume 6 covers the long-neglected subject of transmitters, antennas, and transmission lines. In sum, the full range of the fundamentals of communications electronics is covered in a manner that provides maximum comprehension with a minimum of effort.

To the many people whose thoughts and discussions have contributed to this series, my sincere appreciation. To my wife, Celia, and my daughters, Ruth and Shirley, with whom I would have preferred to spend more time, my heartfelt thanks and gratitude for their assistance and understanding patience.

Falmouth, Massachusetts

BASIC RADIO
vol. 5

The Transistor—Introduction

Semiconductors are not as new as many people believe. The "cat whisker" crystal detector used in the old crystal radios of the early 1900s was actually a semiconductor device. Copper oxide and selenium rectifiers, used in radios as far back as the 1920s, were semiconductor devices also. But, it was the inventor of the transistor at Bell Laboratories in 1948 that brought on the "solid-state" revolution. The transistor was at first a laboratory curiosity, of interest mainly because it was a "solid-state" device that could be made to amplify. Improvements, particularly in the manufacturing process,

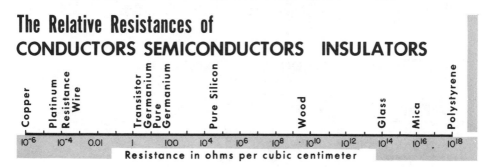

The Relative Resistances of
CONDUCTORS SEMICONDUCTORS INSULATORS

rapidly made it a practical device. Most of today's electronic equipment uses transistors and other semiconductor devices rather than electron tubes; within this decade, they will replace tubes almost entirely, except in the few special cases where tubes are still superior.

In comparison to vacuum tubes, the transistor is extremely small, lightweight, does not require filament power, is virtually immune to mechanical shock and, in general, can operate well at relatively high temperatures. In the following pages, we will discuss semiconductors, transistors, and practical radio circuits containing transistors. We start with a study of semiconductors because this is the "stuff" of which transistors are made.

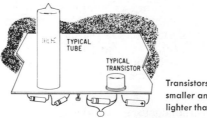

Transistors are much smaller and much lighter than tubes.

A semiconductor material is one whose ability to conduct an electric current falls somewhere between that of an insulator such as mica or glass, and that of a conductor such as silver or copper. Silicon and germanium are the most widely-used materials in the manufacture of transistors. These materials, which in their pure state are very poor conductors, gain important electrical conductivity characteristics with the addition of certain impurities.

Atomic Structure of Semiconductors

An atom of any substance consists of a nucleus made up of neutral particles (neutrons) and positively-charged particles (protons) surrounded by a group of negatively-charged particles (electrons). The electrons travel in orbits around the nucleus. In a <u>neutral atom</u>, there are as many electrons as protons.

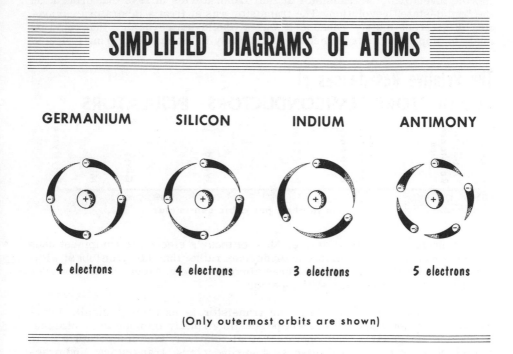

SIMPLIFIED DIAGRAMS OF ATOMS

GERMANIUM   **SILICON**   **INDIUM**   **ANTIMONY**

4 electrons   4 electrons   3 electrons   5 electrons

(Only outermost orbits are shown)

In our study of semiconductors, we will be primarily concerned with the silicon, germanium, indium, and antimony atoms. The silicon atom has a total of 14 electrons in 3 orbits. <u>Four</u> of these electrons are in the outermost orbit. The germanium atom has 32 electrons in 4 orbits. Again, there are four electrons in the outermost orbit. The indium atom contains 49 electrons in 5 orbits, with <u>three</u> electrons in the outermost orbit. Of the 51 electrons in the antimony atom (also arranged in 5 orbits), <u>five</u> are in the outermost orbit. The outermost orbit of these atoms can hold 8 electrons. Thus, none of the above atoms has a complete outer orbit.

In dealing with semiconductors, we are concerned only with the nucleus and the outermost electrons. When shown in the simplified diagram, the atoms of germanium and silicon <u>appear</u> identical because they have the same number of electrons in their outermost orbits. We are concerned primarily with the electrons in the outer ring or orbit (called valence electrons) because they determine the chemical characteristics of the atom or element. Since germanium and silicon have the same number of valence electrons, they have similar chemical characteristics.

Semiconductor Crystals

In a single crystal of pure germanium or silicon, the individual atoms are aligned in a structure which we descriptively call a lattice. The lattice effect is created because all adjacent atoms are symmetrically equidistant. Each atom is held in place by bonds formed between each of its outermost electrons to each of the four adjacent atoms. Each outermost electron thus becomes linked in an electron-pair bond. Thus, a perfect germanium crystal has neither an excess nor a deficiency of electrons. All of the electrons in the crystal are either bound to the nuclei of the atoms or to each other in electron-pair bonds.

From the illustration, it might appear that the electrons in the outer ring can easily be displaced by the application of a voltage. This is not the case. By forming covalent bonds, or electron pairs, between neighboring atoms, the atoms behave in many respects as though their outer ring were complete (8 electrons). This makes them extremely inactive since there are no free electrons; their outer orbits each contain four electrons, and they "share" four additional electrons.

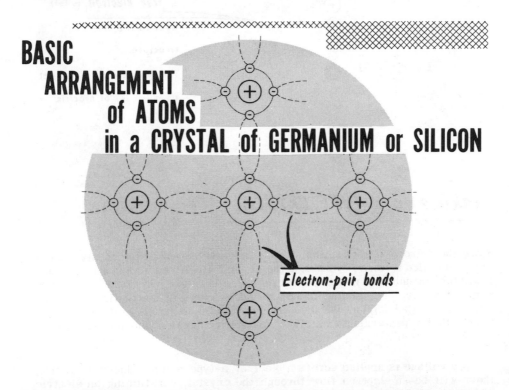

BASIC ARRANGEMENT of ATOMS in a CRYSTAL of GERMANIUM or SILICON

Electron-pair bonds

A perfect germanium or silicon crystal is an electrical insulator and of very little use in transistor work. However, with the addition of a small amount of chemical impurities, heat, or light energy, the conductivity of the crystal is improved and it becomes a semiconductor.

N-Type Semiconductors

If we introduce an impurity such as an atom of antimony (which has <u>five</u> electrons in its outer orbit) into a crystal of germanium or silicon, the crystal lattice as a whole is not changed. The antimony atom merely replaces a germanium or silicon atom, with four of its outermost electrons joining in covalent bonds with those of adjacent atoms. The remaining electron is free to move through the crystal. This free electron aids the conduction of electricity; the antimony-enriched germanium or silicon crystal conducts current more easily than does a pure crystal. Because only small quantities (about 1 part in 100,000,000) of impurities are added, the crystal does not become a <u>good</u> conductor; rather it falls somewhere between a conductor and an insulator. For this reason, these crystals are called <u>semiconductors</u>.

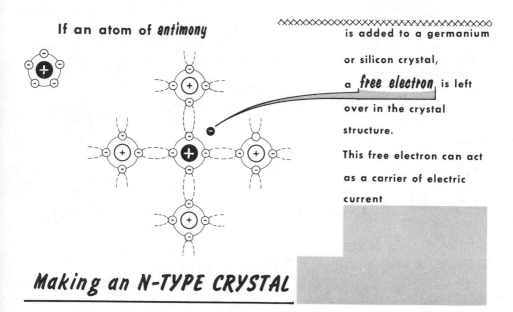

If an atom of *antimony*   is added to a germanium

or silicon crystal,

a ***free electron*** is left

over in the crystal

structure.

This free electron can act

as a carrier of electric

current

Making an N-TYPE CRYSTAL

Since the introduction of antimony <u>adds</u> a free electron to the crystal, antimony is called a <u>donor</u>. Although the crystal itself remains neutral, it has available an unbound negative charge and, therefore, is called an <u>n-type</u> semiconductor. The adding of controlled amounts of impurities to a pure crystal is called <u>doping</u>. Because antimony has 5 electrons in its outer ring, it is called a "pentavalent" material. Another pentavalent substance frequently used is arsenic.

When a voltage is applied across a piece of n-type semiconductor material, there will be an electron flow through the crystal, constituting an electric current. The free electrons will flow toward the positive terminal and be repelled away from the negative terminal. If the voltage were reversed in polarity, all conditions would remain the same except that the direction of current would be reversed. N-type material is a "linear" conductor.

P-Type Semiconductors

If we add a single atom of indium to a crystal of germanium or silicon, a different type of crystal results because indium has only three electrons (trivalent) in its outer orbit. When the indium atom replaces one of the germanium or silicon atoms in the lattice, the crystal becomes deficient in one electron. To fill the vacancy thus created, an electron must be borrowed from some other part of the crystal. The borrowed electron moves into the hole created by adding the indium. In moving to fill the hole, the borrowed electron leaves another hole behind it. Thus, while the electrons move in one direction, the holes left by them move in the opposite direction. Because the loss of an electron (which is a negative charge) results in effect in a positively-charged particle, holes may be thought of as free positive charges that can act as current carriers.

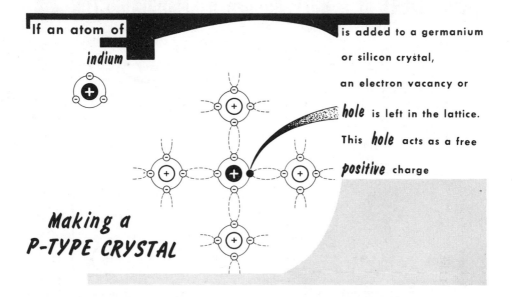

If an atom of indium is added to a germanium or silicon crystal, an electron vacancy or hole is left in the lattice. This hole acts as a free positive charge

Making a
P-TYPE CRYSTAL

Because the indium atom takes an electron from some other atom in the crystal, it is called an acceptor. Since the crystal has a free positive charge available for electrical conduction, it is called a p-type semiconductor. The crystal as a whole remains electrically neutral.

When a voltage is applied across p-type material, the holes, having an effective positive charge, will be attracted to the negative terminal and repelled from the positive terminal. This drift of holes constitutes an electric current, and is equivalent to a flow of electrons in the opposite direction. Each time a hole reaches the negative terminal, an electron is emitted from the negative terminal into the hole in the crystal to neutralize it. At the same time, an electron from a covalent bond enters the positive terminal to leave another hole in the crystal. This hole then begins its drift toward the negative terminal.

Characteristics and Properties of Holes

For an understanding of transistors, we must think of a hole as a specific particle. Holes in motion are the same as electrons in motion--they both make up an electric current. However, we must consider the differences that do exist. A hole can exist only in a semiconductor material such as silicon or germanium because it depends for its existence on a specific arrangement of electrons (electron-pair bonds) and atoms as found in crystal materials. Holes do not exist in conductors such as copper. Another important consideration is that holes are deflected by electric and magnetic fields in exactly the same way as electrons. However, since holes possess a charge equal and opposite to that of the electron, the direction of deflection of the hole is opposite to that of the electron.

MOVEMENT of a HOLE through a SEMICONDUCTOR

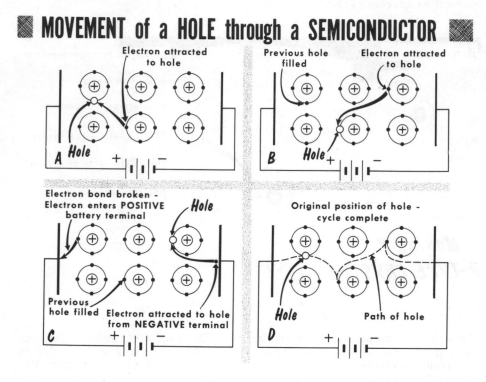

When a hole is filled by a free or excess electron from an adjacent electron-pair bond, the hole no longer exists. When a voltage is applied across a piece of p-type germanium, holes are repelled by the positive terminal and attracted to the negative terminal. When a hole reaches the negative terminal, an electron from the battery (see illustration) enters the germanium and fills the hole. When this occurs, an electron from an electron-pair bond in the crystal near the positive terminal breaks its bond and enters the positive terminal. The breaking of the bond creates another hole which begins to drift toward the negative terminal. This action provides a continuous flow of electrons in the external circuit and holes in the germanium.

The Junction Barrier

If an n-type crystal is joined to a p-type crystal, electrons do not flow across the junction to fill or neutralize the holes. The antimony atoms (actually, ions) in the n-type crystal are positively charged because they have given up electrons. Thus, they repel the positive holes in the p-type crystal. Similarly, the indium atoms in the p-type crystal, in accepting an electron, become negatively charged and repel electrons in the n-type crystal. Free electrons and holes in the n- and p-type material therefore move <u>away</u> from the p-n junction rather than toward it; the junction may be thought of as a barrier to the passage of current carriers. The combined crystal thus acts as if a small equivalent battery were placed across the junction, thereby establishing a small voltage across it. If the current carriers are to pass, the barrier must be eliminated.

A p-n junction cannot be formed merely by placing a piece of n-type material against a piece of p-type material. The junction is formed by taking a single crystal of pure germanium and treating one section with a trivalent impurity and one section with a pentavalent impurity. The semiconductor device we get from this process is called a junction diode. As we shall see, this p-n junction is no longer a linear device. Current will flow much easier in one direction than in the other.

HOLES and ELECTRONS do not have enough energy
to "climb" the *Junction Barrier*, and move away from it

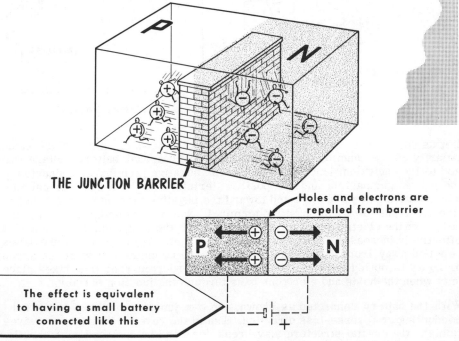

THE JUNCTION BARRIER

Holes and electrons are
repelled from barrier

P   N

The effect is equivalent
to having a small battery
connected like this

−   +

Reverse Biased Junction

Let us review an important point. It might be thought that in time, the electrons in the n-type germanium would, by diffusion, occupy the holes in the
p-type germanium, thereby neutralizing the entire crystal. This does not
occur because the electrons and holes tend to drift apart. In n-type material,
the atoms of the pentavalent impurity have a positive charge; in p-type material, the atoms of the trivalent impurity carry a negative charge. These relatively fixed atoms repel the charges in the opposite piece of material--the
positive atoms in the n-type material repel the holes or positive charges in
the p-type material, and vice-versa. This action provides a battery equiva-
lency across the p-n junction.

REVERSE BIAS RAISES the JUNCTION BARRIER
and NO current flows

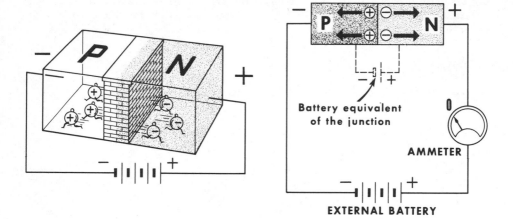

Let us now connect an external battery across our p-n crystal with the same
polarity as the junction-barrier voltage. The external battery voltage will
add to the equivalent junction voltage. The negative terminal is connected to
the p-type germanium and the positive terminal to the n-type germanium.
The positive holes are attracted toward the negative terminal and the negative electrons toward the positive terminal. Note that in both attractions, the
holes and the electrons are attracted away from the p-n junction. This action
effectively increases the junction barrier height. With electrons and holes
repelled away from the junction, there will be no current flow of electrons
or holes through the germanium.   As we will see, current flow takes place
only when the holes and electrons pass through the junction barrier.

With the battery connected as shown, the p-n junction is biased in the non-
conducting or reverse-bias direction. Should the reverse bias be excessive-
ly high, the crystal structure may break down and be damaged permanently.

Forward Biased Junction

We will now connect the positive terminal of the external battery to the p-type
germanium and the negative terminal to the n-type germanium. The holes
are now repelled from the positive terminal of the battery and drift toward
the p-n junction. The electrons are repelled from the negative terminal of
the battery and also drift toward the junction. Under the influence of the battery
voltage, the holes and electrons penetrate the junction and combine with each
other. For each combination of an electron and a hole, an electron from the
negative terminal of the external battery enters the n-type germanium and
drifts toward the junction. Similarly, an electron from an electron-pair bond
in the crystal, near the positive terminal of the external battery, breaks its
bond and enters the positive terminal of the battery. For each electron that
breaks its bond, a hole is created which drifts toward the junction. Recom-
bination around the junction region continues as long as the external battery
is connected.

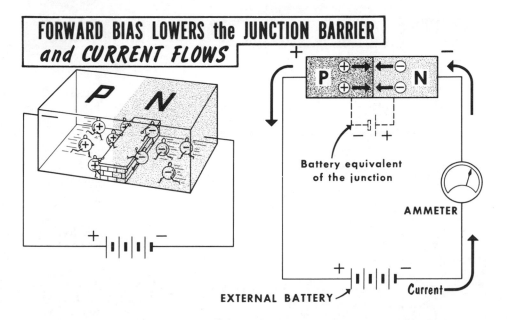

Note that there is a continuous flow of electron current in the external circuit.
The current in the p-type germanium consists of holes; the current in the
n-type germanium consists of electrons. In this condition, the p-n junction
is said to be biased in the forward direction. If the forward bias is increased,
the current is increased.

In forward bias, the external battery voltage opposes and overcomes the
junction-barrier voltage, which may be only a few tenths of a volt. This
lowering of the junction barrier permits a free flow of current. Of course,
excessive forward bias would produce excessive current, with a possibility
of a crystal-structure breakdown.

The Junction Diode as a Rectifier

We have seen that the p-n junction is a unilateral device; that is, when forward biased, it will permit current to flow, and when reverse biased, it will
not permit current to flow. These, then, are the basic ingredients of a diode.
Below, we see a curve showing current flow through a junction diode as the
bias voltage is varied and reversed in polarity. Note that current flow in the
forward-bias direction is quite high--measured in milliamperes. However,
current flow in the reverse-bias direction, although very low and measured
in microamperes, is not zero. The reverse-bias current flow occurs because some acceptor ions and their associated holes occur in the n-type germanium, and some donor ions and their associated excess electrons occur in
the p-type germanium. The holes found in n-type germanium and the excess
electrons in the p-type germanium are called minority carriers because they
are so few in number, compared with the holes found in p-type material and
the electrons in n-type germanium, which are called majority carriers.

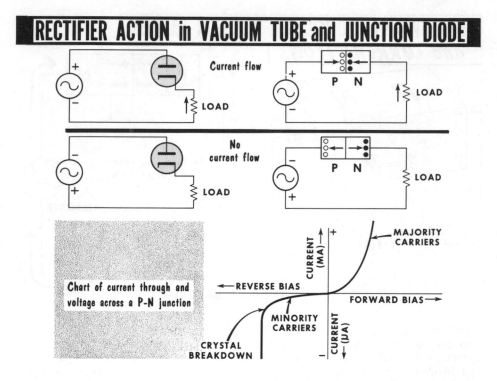

Note that when a very high reverse bias is applied, a high reverse current
flows.   This current is not due to the minority carriers, but to a breakdown
of the crystal structure. The point at which the reverse voltage is high enough
to break down covalent bonds and cause current flow is called the Zener breakdown voltage.   This voltage has the same importance as the inverse voltage
rating of a vacuum tube, since it defines the maximum reverse voltage that
can be applied to a junction without excessive current flow.

The Point-Contact Diode

Up until this point, we have given all our attention to the junction diode. There is an older type called the point-contact diode. This unit is an outgrowth of the old galena crystal detector, which consisted of a piece of galena (a lead ore) mounted so that an irregular surface was exposed to the point of a wire (cat whisker). The wire was moved around to find a sensitive spot for optimum radio detection.

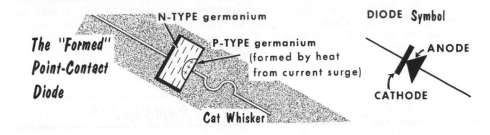

The modern counterpart of this is the germanium crystal point-contact diode. Here, the germanium replaces the galena, and the cat whisker consists of a length of wire such as tungsten about 0.005" diameter. In a modern unit, the cat whisker is fitted to the germanium crystal at the factory, and the entire unit is sealed. In practice, the germanium consists of n-type material. The unit is "formed" by passing a large momentary surge of current across the junction of the wafer and the whisker. The heat produced by this current forces some electrons away from the area of the point, leaving holes. This produces a small p-type region around the point of the cat whisker.

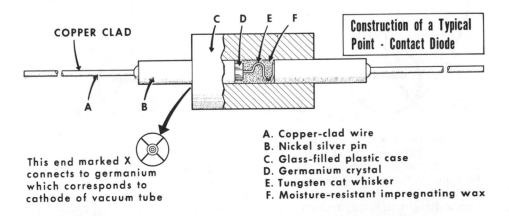

This end marked X connects to germanium which corresponds to cathode of vacuum tube

A. Copper-clad wire
B. Nickel silver pin
C. Glass-filled plastic case
D. Germanium crystal
E. Tungsten cat whisker
F. Moisture-resistant impregnating wax

Thus, we have all the ingredients of a junction diode--a p- and an n-region. However, since the p region is so tiny, there is very little capacitance across the junction. A typical shunt capacitance might be as low as 0.8 μpf. This makes the point-contact diode highly desirable for high-frequency work such as in video detectors and microwave mixers.

What is a Transistor?

So far, we have discussed p- and n-type materials and their actions. Going one step further, we have observed the action of p-n junctions under conditions of forward and reverse bias. We are now prepared to study the transistor. We can consider a junction transistor as being composed of two separate p-n junctions "tied" together. One basic type of junction transistor is known as the p-n-p. In this transistor, there is a very thin layer of n-type semiconductor between two much thicker layers of p-type semiconductor material. The second basic type of junction transistor is the n-p-n. Here, the opposite condition exists. A thin layer of p-type semiconductor material is between two much thicker layers of n-type material. As we will see, it is very important that the center layer of semiconductor material be made extremely thin.

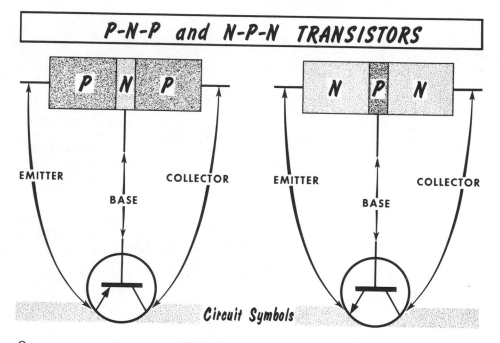

P-N-P and N-P-N TRANSISTORS

EMITTER   COLLECTOR   EMITTER   COLLECTOR

BASE   BASE

Circuit Symbols

Once a p-n-p or n-p-n junction transistor is formed, electrical connections are made to the layers. The connection to the thin center layer is called the base, and can be compared roughly to the control grid of a vacuum tube. The connection to one of the outer layers is called the emitter, often compared to the cathode of a vacuum tube. The connection to the other outer layer is called the collector, often compared to the plate of a vacuum tube. In some transistors, the outer layers are symmetrical and can be used either as the emitter or collector, depending upon circuit biasing voltages. However, it is much more common practice to make the junctions asymmetrical, and the manufacturer will identify each lead as being either the emitter, the base, or the collector. The illustration shows the physical representation of p-n-p and n-p-n transistors together with their circuit symbols.

Transistor Basing Diagrams and Construction

Several transistor outline drawings are shown on this page to illustrate the various physical shapes to be found in the transistor field, together with the location of the emitter, base, and collector leads. Among the typical transistors shown, note that the diameter may be as little as 0. 322", or less than one-third inch. The leads are generally thin, tinned wires that must be handled carefully. On some large power transistors (to be discussed later), the collector may be connected to the case and the entire unit bolted to the chassis.

Viewed from the Side or Bottom, the *TYPICAL TRANSISTOR* Looks Like This

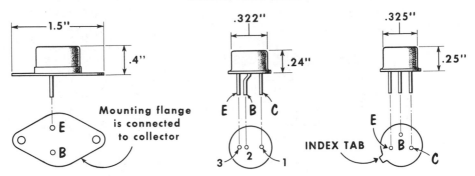

To identify the leads, an index or color dot is often placed next to the collector terminal. Sometimes, a small index tab is used. It is always good practice to check the manufacturer's schematic for transistor lead identification before any testing is done.

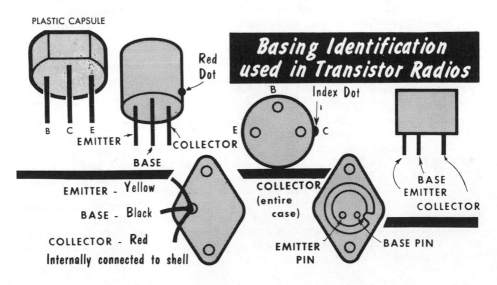

Operation of the P-N-P Transistor

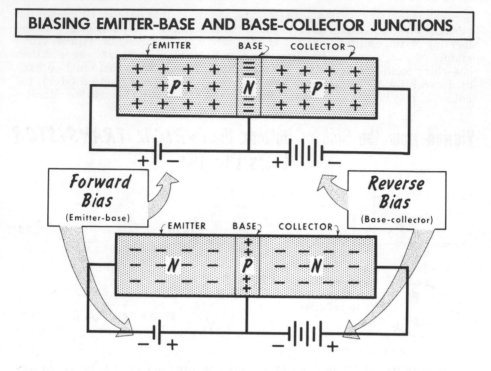

BIASING EMITTER-BASE AND BASE-COLLECTOR JUNCTIONS

Let us observe the action of a p-n-p transistor. In this unit, there are two p-type sections, the emitter and collector, separated by a thin layer of n-type material, the base. Note that the emitter-base p-n junction is biased in a forward direction. In this connection, the external battery opposes the internal (barrier junction) voltage developed at this junction. In short, with the emitter biased positively with respect to the base, we say this p-n junction is forward biased. The p-n collector-base junction is biased in the opposite direction. In this way, the external battery aids the internal junction voltage, and we say that the collector is biased negatively (reverse bias) with respect to the base.

If no voltage were applied to this transistor, the holes in the emitter would move to the left and the electrons in the base would move to the right, both because of the internal junction barrier voltage. However, when an emitter-base forward-bias voltage is applied, the holes in the emitter and the electrons in the base move toward the junction, both moving in the direction of the battery terminal which attracts them. At the emitter-base junction, some holes and electrons combine with each other and are neutralized. However, because of the extreme thinness of the base layer, and because of the attraction of the relatively high negative collector voltage, almost all of the holes pass or "diffuse" through the base and produce a "hole current" between the emitter and the collector.

Operation of the P-N-P Transistor (Cont'd)

The amount of holes combining with electrons in the emitter-base junction region is quite small--usually less than 5%. However, because of this combination, there is a small emitter-base current, and the collector current is less than the emitter current by that amount. For example, if the emitter current is 1 ma, the collector current would be about 0.95 ma. Holes reaching the negative battery terminal at the collector combine with electrons from the battery. At the same time, new holes are formed at the emitter by electrons breaking their electron bonds and entering the positive battery terminal. Thus, while the current carriers within the transistor consist of holes, the external current consists of electrons.

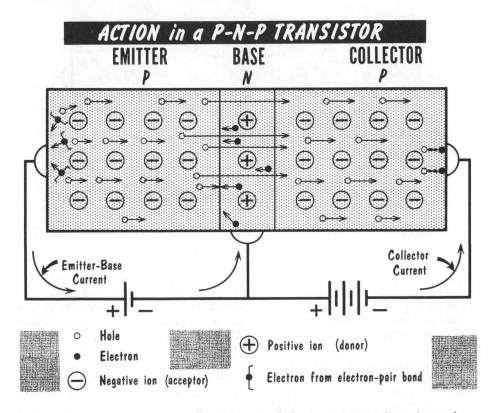

As we shall see, maximum effectiveness of the transistor takes place when all the holes or current carriers from the emitter pass through the base region and on to the collector. Loss of some of these current carriers is unavoidable because of the electrons in the base region. However, by making the base sufficiently thin, the percentage of the holes that diffuse through without combining with electrons may reach as high as 99%. The holes that combine with base electrons set up an "undesirable" emitter-base current, with electrons leaving the emitter to flow into the emitter-base battery, and electrons leaving the battery to flow into the base.

Operation of the N-P-N Transistor

The functioning of an n-p-n transistor is very similar to that of the p-n-p type. The important difference is that the current carriers are now <u>electrons</u> instead of holes. Here again, the internal junction barrier voltage in the emitter-base region is offset by the externally applied bias voltage which forces the electrons to the right and the holes to the left. The electrons from the emitter

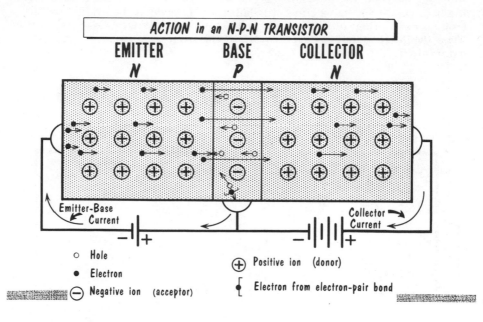

ACTION in an N-P-N TRANSISTOR

enter the center or base region, which has a p-type conductivity, and here, a small percentage of them combine with holes.   Those electrons that do not combine with holes pass, or diffuse through the base region into the n-type collector under the influence of the high positive collector voltage.   These electrons constitute the collector current.

Since a certain number of electrons combine with holes at the emitter-base junction, the collector current will be somewhat less than the emitter current.   Actually, the amount of emitter current that enters into combination is usually less than 5% of the total current. Generally, the collector current is proportional to the collector voltage for small collector voltages.   However, for a given emitter current, the number of current carriers is constant; above a certain collector voltage, the collector current is nearly constant.

The fundamental differences between n-p-n and p-n-p operation are as follows: The emitter-to-collector current carrier in the p-n-p transistor is the hole; in the n-p-n transistor, the emitter-to-collector current carrier is the electron.   Also, the bias voltage polarities are reversed, resulting from a reversal of n-type and p-type semiconductor materials.

Collector Voltage--Collector Current Characteristics

If we connect a simple transistor circuit as shown, we can vary the collector voltage while observing its effect on the collector current. As we start from zero collector voltage and increase it, there is a rapid and constant increase in the movement of free current carriers, with a resultant increase in collector current. A further increase in collector voltage will produce still more current carrier activity with a greater increase in collector current. However, a point soon is reached where a still further increase in collector voltage will result in very little change in collector current. This condition can be compared with that of plate current saturation in a vacuum tube. Actually, in transistors, a point is reached where the number of free current carriers available (electrons in n-type material and holes in p-type material) is no longer great enough to permit a significant increase in collector current. From this point on, the curve is almost horizontal.

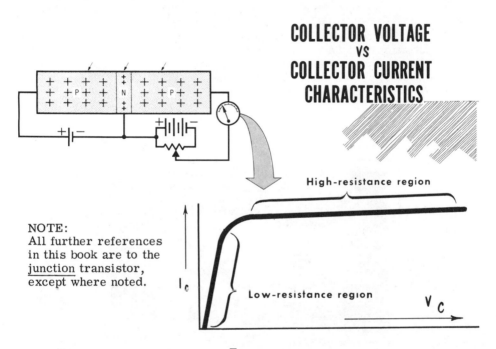

COLLECTOR VOLTAGE
VS
COLLECTOR CURRENT
CHARACTERISTICS

High-resistance region

NOTE:
All further references
in this book are to the
junction transistor,
except where noted.

I_c

Low-resistance region

V_c

We know from Ohm's law that $R = \dfrac{E}{I}$ We learned also that the plate resistance of a tube can be measured by E_p/I_p. We can therefore find the collector resistance by V_c/I_c. (In transistor terminology, voltage is often expressed as "V" rather than "E.") In the very early portion of the curve, the rise is very great, indicating a low collector resistance. After the "bend," there is very little change in collector current for a given change in collector voltage, indicating a very high collector resistance. This is similar to the high plate resistance of the pentode. In practice, the collector is almost always biased in this region of high resistance. The resistance in this area may be as high as 1 or 2 megohms.

Current Amplification Factor--Alpha

The current amplification of a transistor can be compared with the voltage amplification factor (μ) of a vacuum tube. This, as you recall, is a measure of the relative ability of the grid voltage and the plate voltage to produce an equal change in the plate current ($\Delta ep / \Delta eg$). Note that the voltage in the output circuit is directly related to the input voltage. For this reason, vacuum tubes are considered as voltage-operated devices. Similarly, it is character- istic of transistors that the current flowing in the output circuit is directly related to the input current. Therefore, transistors are considered as cur- rent-operated devices. Thus, in discussing transistors, we think in terms of input and output current. This does not mean that transistors cannot be used as voltage amplifiers. Actually, they are, and we will see how voltage gain is obtained in transistor amplifiers.

The change in collector current caused by the change in emitter current, assuming a constant collector voltage, is called the current gain of a tran- sistor. This current gain, or alpha (α), is expressed: $\alpha = i_c / i_e$. Since every transistor will have some emitter-base current, alpha cannot reach unity or 1 in a conventional junction transistor. However, alphas of between 0.95 and 0.99 are obtainable. As we will see, current gain (α) applies pri- marily to common-base amplifiers.

Another item of interest related to transistors is resistance gain. With the emitter-base circuit biased in the forward direction, the internal resistance of the input circuit is low. When the base-collector circuit is biased in the reverse direction, it has, as we have seen, a very high internal resistance. As we shall see, this resistance gain is very important. The resistance gain of a transistor is expressed as the ratio of the emitter-base internal resis- tance divided by the collector-base internal resistance. Said in another way, it is the output resistance divided by the input resistance (RG = $\frac{R_o}{R_i}$).

Transistor Voltage Gain

In our study of current gain (alpha), we learned that any change in emitter current produces a change in collector current alpha times as great. Thus, in a junction transistor, current "amplification" is generally limited to about 0.95 or so. Therefore, current "gain" in the usual sense is not obtainable. However, voltage gain is. For purposes of explanation, let us consider a junction transistor having a current gain of unity, or 1. Thus, any a-c component in the emitter circuit will produce the same a-c current fluctuation in the collector circuit.

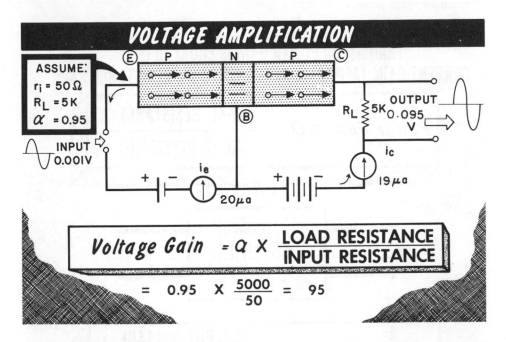

Let us now assume that an input signal voltage of 1 millivolt is applied to the input (emitter-base) circuit. Because of the forward bias, the input circuit resistance is low--say 50 ohms. Now we can compute the emitter current (i_e) flow, using Ohm's law: i_e = .001/50, or 20 microamperes. Since the current gain of this transistor is 1, the collector current i_c also would be equal to 20 μa. However, because of the reverse bias, the collector circuit has a very high internal resistance. This high resistance is important because it permits a high-value load resistor to be used in the collector circuit without affecting the output current. This, then, is the key to voltage amplification in transistor circuits. Since the output voltage across the load resistor is equal to the output current times the load resistance, say 5000 ohms, the output voltage is equal to 20 μa times 5000 ohms, or 0.1 volt. Since the input voltage was 0.001 volt, this represents a voltage gain of 100. Thus, we can sum up voltage gain by saying it is equal to alpha times the ratio of load resistance to input resistance. In the above example, if alpha was 0.95, voltage gain would have been 0.95 × 100, or 95.

Transistor Power Gain

From our discussion of voltage gain in a transistor, we can move easily to an understanding of power gain. The input power delivered to the input circuit of a transistor can be obtained from Ohm's law formula, $P = E^2/R$. In terms of our transistor circuit, this would be equal to the input voltage squared divided by the input resistance (the resistance of the emitter-base circuit), or V_{in}^2/r_i. The power delivered to the external collector load circuit would be equal to the output voltage squared divided by the load resistance, or V_{out}^2/R_L. The power gain of the amplifier is, then, the ratio of output power divided by input power.

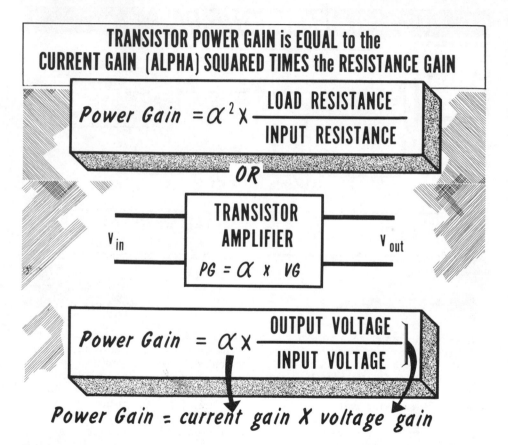

TRANSISTOR POWER GAIN is EQUAL to the CURRENT GAIN (ALPHA) SQUARED TIMES the RESISTANCE GAIN

$$Power\ Gain = \alpha^2 \times \frac{LOAD\ RESISTANCE}{INPUT\ RESISTANCE}$$

OR

TRANSISTOR AMPLIFIER

V_{in} V_{out}

$PG = \alpha \times VG$

$$Power\ Gain = \alpha \times \frac{OUTPUT\ VOLTAGE}{INPUT\ VOLTAGE}$$

Power Gain = current gain X voltage gain

Since voltage gain is equal to α times R_L/R_i, it follows that power gain would be equal to α^2 times R_L/R_i, since power is also equal to I^2R. Simplifying it still further, we can state that power gain is equal to α times V_{out}/V_{in}. This is reasonable, for if the voltage gain of an amplifier is the ratio of V_{out}/V_{in}, and if the current gain is equal to α, the power gain must be the product of these two; that is, $\alpha\ V_{out}/V_{in}$. Since the power gain in a junction transistor is equal to alpha times the voltage gain, we can see that the power gain will always be slightly less than the voltage gain.

Transistor Frequency Response--Alpha Cutoff

The time it takes electrons and holes to pass from the emitter to the collector (input to output circuit) is called the transit time. The transit time of these current carriers is one of the major factors limiting the high-frequency response of transistors. The movement of holes or electrons from the emitter through the base layer to the collector requires a short but finite time. In the transistor, the electron does not have a clear or unimpeded path from emitter to collector. As a result, the transit time is not the same for all electrons injected into the emitter at any one instant. If the transit time for all electrons were the same, there would be a simple delay in time between the input and output signals. Since the injected carriers do not all take the same path through the transistor, those produced by a single pulse at the emitter do not all arrive at the collector at the same time. The resulting difference is very small and unimportant in the audio-frequency stage. However, at the higher frequencies, the difference becomes a significant part of an operating cycle and causes partial cancellation between carriers. This causes a reduction in amplitude of the higher frequencies. The decrease in the output signal means a decrease in alpha, the current gain. In addition, the degradation in frequency response becomes steadily worse as the operating frequency is increased, until eventually there is no relationship, and no gain, between input and output waveforms.

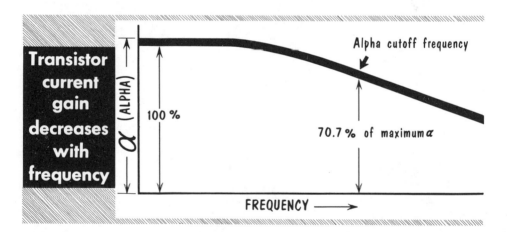

Another limitation of high-frequency response is the capacitance between transistor regions, as well as between transistor leads. As the frequency increases, capacitive reactance decreases, and there is a tendency for signals to be shunted across the emitter-collector circuits.

The alpha cutoff frequency is the frequency at which current amplification has fallen to 0.707, or a loss of 3 db, from its low-frequency value, usually measured at 1 kHz or lower. The alpha cutoff frequency is generally considered as the highest "useful" frequency amplified by a transistor in a common-base circuit. It is determined largely by the size of the emitter and collector electrodes, and the thickness of the base region.

Transistor Circuit Diagrams

So far, basic transistor circuits have been shown in a manner similar to the transistor's physical construction. Now, however, we will begin working with schematic diagrams of circuits using conventional transistor symbols. These symbols are shown below, along with simple schematic diagrams of transistor circuits.   Note particularly that on the symbol for an NPN transistor the arrow points <u>outward,</u> and on the symbol for a PNP transistor it points <u>inward.</u> Note also that the polarity of the biasing batteries is exactly the opposite for NPN and PNP transistors.

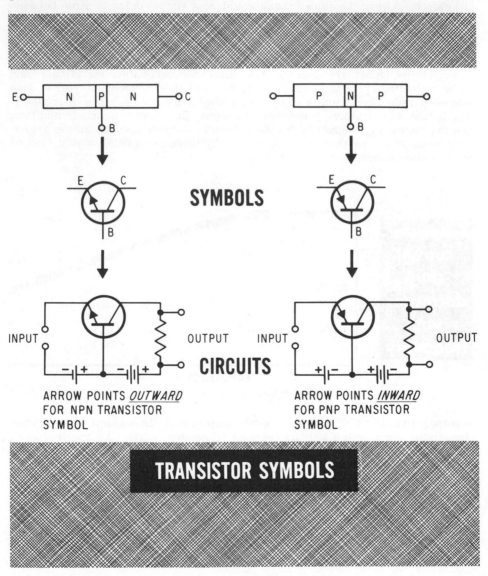

SYMBOLS

CIRCUITS

ARROW POINTS *OUTWARD*
FOR NPN TRANSISTOR
SYMBOL

ARROW POINTS *INWARD*
FOR PNP TRANSISTOR
SYMBOL

TRANSISTOR SYMBOLS

The smallest part of an element which can take part in chemical changes is called an atom.

Atoms consist of positively-charged particles called protons, negatively-charged particles called electrons, and uncharged particles called neutrons.

A semiconductor has a resistivity between that of a conductor and an insulator. Examples of semiconductors are germanium and silicon.

In a crystal, the atoms are arranged in a specific pattern called a lattice.

Electrons shared by adjacent atoms in a crystal form electron-pair bonds (covalent bonds).

N-type germanium contains donor impurities--materials having five valence electrons. One of these electrons cannot form an electron-pair bond, and is called an excess electron.

Common donor materials include arsenic, antimony, and boron.

P-type germanium contains acceptor impurities--materials having three valence electrons. Since four valence electrons are needed to complete all adjacent electron-pair bonds, a hole is created.

A hole can be considered a positive charge which diffuses or drifts through a crystal. The drift of holes constitutes a current.

Forward bias of a p-n junction causes heavy current (flow of majority carriers). Reverse bias causes very low current (flow of minority carriers).

Holes constitute the principal current through the p-n-p transistor. Electrons constitute the principal current through the n-p-n transistor.

The emitter, base, and collector of the transistor are comparable to the cathode, grid, and plate of the vacuum tube.

The emitter-base junction is normally biased in the forward (low resistance) direction.

The base-collector junction is normally biased in the reverse (high resistance) direction.

Collector current depends upon the emission of carriers from the emitter-base barrier.

Alpha (α) is an expression of current amplification in the common-base amplifier. It is a measure of a change in collector current to a change in emitter current, with the collector voltage kept constant.

The alpha cutoff frequency indicates the point where the gain of a transistor has fallen to 0.707 of its maximum gain.

REVIEW QUESTIONS

1. What is meant by covalent bonds?
2. Explain the differences between n-type and p-type semiconductors.
3. What is a junction barrier?
4. Explain the operation of a p-n junction during conditions of forward bias and reverse bias.
5. How does a junction diode act as a rectifier?
6. Explain the basic construction of a transistor.
7. Explain the operation of p-n-p and n-p-n transistors.
8. What is the meaning of alpha and what is its significance?
9. How is a voltage gain acheived in a junction transistor circuit?
10. What is meant by the alpha cutoff frequency?

Three Basic Transistor Circuits

The three-element transistor we have been studying (p-n-p or n-p-n) can be connected into three basic type circuits--common base, common emitter, and common collector. Since there are only three connections to a transistor, and since each transistor circuit must have an input circuit requiring two input leads and an output circuit requiring two output leads, it follows that one of the three transistor leads must be common for both the input and output circuits. Very often, the common lead is used as a reference point for the entire circuit, and is thus connected to chassis or ground. This gives rise to the expressions grounded base, grounded emitter, and grounded collector. Common and grounded mean the same thing. In some instances, the element-- emitter, base, or collector--will be connected directly to ground. Where this occurs, the element is at both a-c and d-c ground. Where the element goes to ground through a battery or resistor that is bypassed by a capacitor, the element is at a-c ground only.

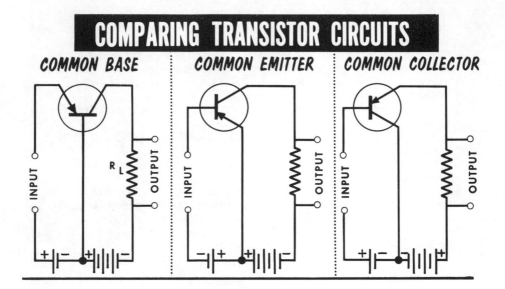

Each circuit has specific characteristics that make it useful in different circumstances. For example, the common base circuit has a low input resistance and a high output resistance, making it suitable for matching a low impedance device to a high impedance circuit. The common collector circuit is the reverse, being a high input resistance and a low output resistance circuit. The characteristics of each of the three circuits will be discussed in the following pages, followed by a comparison of the different type circuits. These characteristics include input and output resistances, and voltage, current, and power gain.

The Common-Base Amplifier

We will begin our study of transistor amplifiers with the common-base circuit, since it is this configuration we are most familiar with, having used it from the beginning of this book. With the emitter-base circuit forward biased and the collector-base circuit reverse biased, current will flow as shown in the diagram. Using an n-p-n transistor, note that the collector current is 95% of the emitter current, with the remaining 5% flowing in the base circuit as a result of electrons from the emitter combining with holes in the p-type base. The details of transistor action have already been discussed.

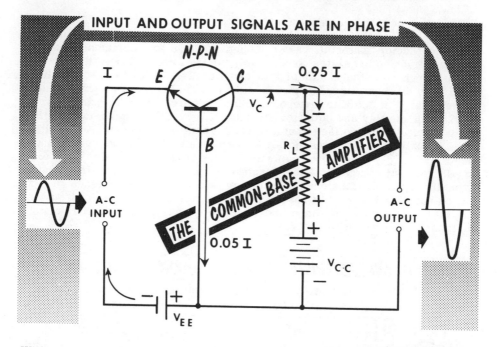

With no signal input, a certain collector current will flow, producing a voltage drop across R_L. This voltage drop is in opposition to the collector battery voltage V_{CC}, and places the collector voltage V_C at some value lower than the battery voltage. Now, we inject a signal into the emitter-base circuit. We will assume that the first half cycle is positive-going and the second half cycle negative-going. Since the emitter is negative with reference to the base, the positive-going half cycle will oppose the negative bias voltage. With a reduction in forward bias, the emitter current will be reduced with consequent resulting reduction of collector current. Since the collector current is reduced, the voltage drop across R_L is reduced. We said that the collector voltage V_C is equal to the battery voltage V_{CC} minus the voltage drop across R_L. Thus, since the voltage drop across R_L is decreased, the collector will become more positive. Hence, we see an important point: as the input cycle varies through its positive half cycle, the output signal developed at the collector also varies through a positive half cycle.

The Common-Base Amplifier (Cont'd)

Now we observe the input signal as it goes through its negative half cycle. As the emitter goes negative, forward bias is increased, collector current is increased, and the voltage drop across R_L is increased. Since this voltage opposes the positive potential of V_{CC}, the collector goes more "negative," or less positive. We can thus conclude: <u>In a common-base circuit, the input and output voltages are in phase</u>--there is no phase reversal. This is exactly the same relationship that exists in the grounded or common-grid vacuum tube circuit.

As we have seen, the current gain (α) of the common-base amplifier is low, always being less than unity. However, by proper design of the center layer of semiconductor material, values as high as 0.98 are commonly reached. A factor which compensates considerably for the low current gain is the extremely high resistance gain of this circuit. The common-base amplifier has a very low input impedance of from 30 to 150 ohms, and a very high output impedance of 300 K to 1 megohm. Since approximately the same current flows in the emitter and collector circuits, a very high load resistance (R_L) can be placed in the collector output circuit, resulting in considerable voltage gain. Actually, voltage gains up to 1000 are not unusual in this circuit arrangement. Relatively good power gains are also available, with gains of 20 to 30 db (100 to 1000) common. A disadvantage of the common-base circuit is the difficulty involved in matching impedances because of the extremes in input and output resistances. However, it is ideal for applications such as amplifying the output of a low-impedance magnetic phono pickup, where no matching transformer would be required.

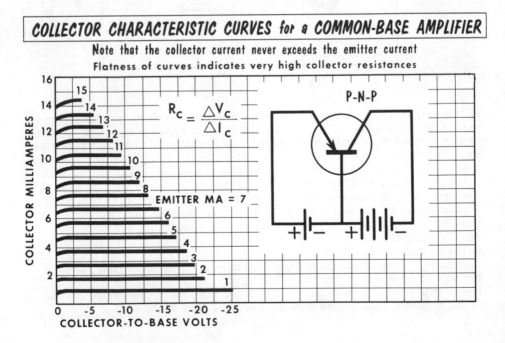

COLLECTOR CHARACTERISTIC CURVES for a COMMON-BASE AMPLIFIER

Note that the collector current never exceeds the emitter current

Flatness of curves indicates very high collector resistances

$$R_C = \frac{\triangle V_C}{\triangle I_C}$$

P-N-P

EMITTER MA = 7

COLLECTOR MILLIAMPERES

COLLECTOR-TO-BASE VOLTS

The Common-Emitter Amplifier

In the common-emitter circuit, once again we have a forward-biased base-emitter circuit and a reverse-biased collector-emitter circuit. Note that the emitter lead is now common to both the input and output circuits. Using the same n-p-n transistor circuit, the current in the collector circuit is again 95% of the emitter current, with the balance flowing in the base circuit.

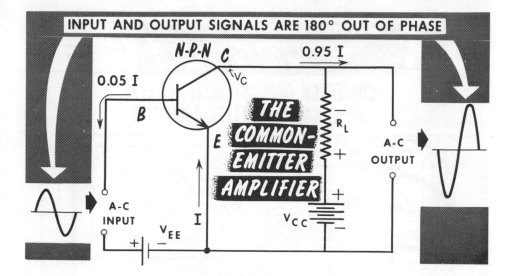

Applying the same input signal, note how the positive half cycle aids the input circuit forward bias. As this positive-going half cycle is applied between base and emitter, it increases forward bias with a resulting increase in collector current. This produces an increased voltage drop across R_L, which subtracts from V_{CC}, making the collector voltage V_C less positive (more negative).

During the negative half cycle, the input signal opposes the forward bias of the input circuit, thereby reducing the emitter and collector current. With a drop in collector current, the voltage drop across output load resistor R_L decreases, making the collector more positive. We now see the phase relationship in the common-emitter circuit--the input and output voltages are 180º out of phase. This is exactly the same relationship that exists in the common-cathode vacuum tube circuit.

The input and output impedances of the common-emitter circuit are considerably less severe than that of the common-base circuit, and impedance matching is much simpler. For instance, the input resistance to the common-emitter circuit may range from 500 to 1500 ohms, while the output resistance is usually in the range of 50 K ohms. Since the resistance gain of the common-emitter is so much less than that of the common-base circuit, it would seem that the available voltage gain would be much less. This is not the case.

The Common-Emitter Amplifier (Cont'd)

In the common-emitter circuit, current gain is measured by control of the collector current by the base current; the term for current gain in this circuit is beta (β). The beta is equal to a change in collector current divided by a change in base current, or β = delta I_c/delta I_b. In its more common form, the current gain of the common-emitter circuit is stated in terms of alpha; that is, beta = alpha/(1 - alpha).

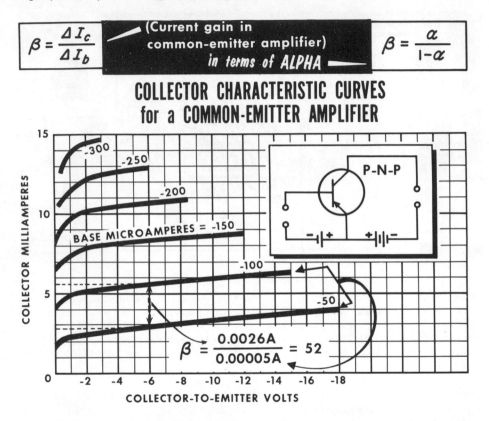

From this, we can see that extremely large beta gains are possible when the alpha characteristic of a transistor is high. For instance, the beta current gain of a common-emitter transistor when alpha is 0.95 is 0.95/(1 - 0.95), or 19. If alpha is as high as 0.98, then the beta current gain is 0.98/(1 - 0.98), or 49. Since an alpha of 0.98 is somewhat high, the current gain of most common-emitter amplifiers is in the range of 35, with values up to 60 attainable. Because of this very high current gain, and even though the resistance gain is very low compared to the common-base amplifier, the voltage gain is still quite respectable, being usually slightly lower than that of the common base. Common-emitter voltage gains of 250 to 500 are attainable. Since power gain is a function of the current squared, very high power gains are possible in this high-current gain circuit. Power gains of up to 40 db (10,000) are common.

The Common-Collector Amplifier

In the common-collector circuit, the collector lead is common to both the input circuit (base-collector) and the output circuit (emitter-collector). Note that the load resistance R_L is in the emitter lead and the output is taken from the emitter. You will recall this being the same situation that exists in the vacuum tube cathode-follower circuit, thus this circuit is often referred to as an "emitter-follower" circuit. In our circuit, using an n-p-n transistor, when we apply a positive-going signal, the input circuit forward bias

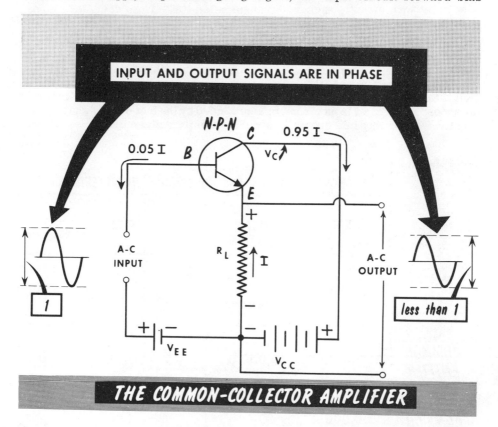

INPUT AND OUTPUT SIGNALS ARE IN PHASE

THE COMMON-COLLECTOR AMPLIFIER

is increased, resulting in an increase in collector current. This current flows through R_L in the emitter circuit, making the emitter more positive. When the incoming signal is negative going, the input circuit forward bias is reduced, lowering the emitter and collector currents. This reduces the voltage drop across the emitter or load resistor, making the emitter negative going, or less positive. Since the output signal is taken across the load resistor between emitter and ground (or common), we see that the output voltage will vary in step with the input signal. From this, we make the observation: In the common-collector circuit, the input and output voltages are in phase. This, of course, is exactly the same situation as exists in the common-plate or cathode-follower circuit.

The Common-Collector Amplifier (Cont'd)

The common-collector circuit is interesting in that it is so different from the two previous circuits. The input resistance of the common-collector circuit is extremely high, ranging from 100 K to 500 K ohms, while the output resistance is low, in the range of 50 to 1000 ohms. The input resistance is high due to the high resistance of the base-collector circuit; the output resistance is low due to the low resistance of the emitter-collector circuit. This circuit makes an excellent isolation network between a high-impedance and a low-impedance circuit, such as a transmission line. In so doing, of course, it is acting as an impedance-matching device.

As in the case of the vacuum tube cathode follower, the voltage gain of the common-collector amplifier will always be less than unity (1), and no real gain is obtained. This is due to the degenerative effect of the load in the emitter circuit. Since the input and output signals are in phase, the output signal acts to oppose changes in input circuit bias produced by the incoming signal. Practically, we can say this circuit is capable of unity voltage gain.

SUMMARY OF TYPICAL TRANSISTOR CIRCUIT CHARACTERISTICS					
	INPUT IMPEDANCE	OUTPUT IMPEDANCE	CURRENT GAIN	VOLTAGE GAIN	POWER GAIN
COMMON BASE	30-150 Ω	300,000 Ω - 1 Megohm	Less than 1	300 - 1000	Up to 1000
COMMON EMITTER	500-1500 Ω	50,000 Ω	35	250- 500	Up to 10,000
COMMON COLLECTOR	100,000 - 500,000 Ω	50- 1000 Ω	35	Less than 1	Up to 1000

The value of current gain for a common-collector amplifier is not very different from that for a common-emitter amplifier. When alpha is nearly unity, $\alpha/(1-\alpha)$ is not very different from $1/(1-\alpha)$. In spite of this high current gain, however, the common-collector gives less than unity voltage gain because the input resistance is so high compared with the load resistance. Current gains of 35 or so are commonplace. The maximum power gain of this circuit is about 30 db.

Transistor Biasing

All transistor circuits discussed so far have used two voltage sources (batteries)--one for the emitter-base bias and one for the base-collector bias. Obviously, these two bias voltages are necessary. However, there are practical means by which we can eliminate the need for two batteries. In doing so, we must be careful that our biasing methods provide the same voltage stabilization provided by batteries. In the comparatively seldom-used common-base amplifier, a simple voltage divider can be used to establish emitter-base bias. Two resistors connected across the collector supply will provide the necessary voltage distribution. Note that with the p-n-p transistor, the voltage drop is of such a polarity as to make the base negative with respect to the emitter; this is as it should be for forward bias. A disadvantage of this type of bias is its inefficiency. The constant voltage-divider bleeder current represents a waste of power.

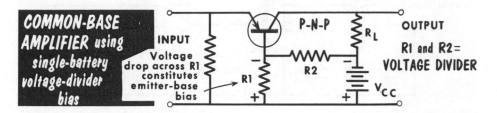

COMMON-BASE AMPLIFIER using single-battery voltage-divider bias

INPUT
Voltage drop across R1 constitutes emitter-base bias

P-N-P

OUTPUT
R1 and R2 = VOLTAGE DIVIDER

The extremely popular common-emitter circuit requires more consideration in the design of its bias arrangement. One method, often called <u>fixed base-current bias</u>, has the base resistor connected directly between the base and the "high" side of the collector battery. Base-bias current will flow in this circuit through Rb and produce a voltage drop opposite in polarity to that of the collector supply; the difference is applied between the emitter and base to provide the necessary input forward bias.

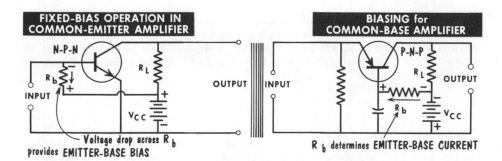

FIXED-BIAS OPERATION IN COMMON-EMITTER AMPLIFIER

N-P-N

INPUT

OUTPUT

Voltage drop across R$_b$ provides EMITTER-BASE BIAS

BIASING for COMMON-BASE AMPLIFIER

INPUT

P-N-P

OUTPUT

R$_b$ determines EMITTER-BASE CURRENT

Another biasing arrangement for common-base amplifiers involves the placing of a resistor in the base circuit between the base and the high side of the collector supply. This biases the emitter positively with respect to the base, and establishes the proper emitter current. Normally, there would be signal degeneration as a result of this resistor, but this is avoided by placing the base at a-c ground potential through a bypass capacitor.

Transistor Biasing (Cont'd)

The connection of the fixed-bias circuit permits a degree of instability to enter, in that the bias voltage changes with changes in ambient temperature which in turn cause transistor current variations. To improve this condition, we use a circuit called <u>self bias</u>. Here, the base resistor is connected directly between base and collector. In this circuit, we improve stability by introducing degeneration (negative feedback) similar to that produced by an unbypassed cathode resistor in vacuum tube circuits. This degeneration, in transistor circuits, is a form of automatic control of the base bias. In this

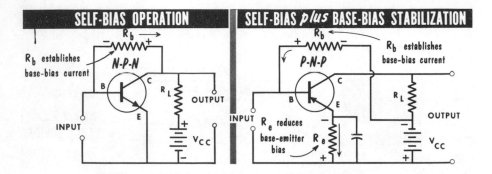

circuit, base-bias resistor R_b establishes the value of collector current and prevents excessive shifts in the collector d-c operating point due to temperature change or transistor replacement. In the self-bias circuit, a change in temperature may cause an increase in collector current. This would cause an increase in voltage drop across the load resistor, reducing the collector voltages. This, in turn, would produce a decrease in the base-bias current, thus compensating for the change. Should there be a decrease in collector current, causing a rise in collector voltage, base-bias current would increase.

The self-bias circuit offers much higher stability than the fixed-bias arrangement. However, the self-bias circuit produces a-c feedback through the bias network which reduces the gain slightly. This feedback is often reduced by using two series resistors in place of R_b, and bypassing the tap to ground through a capacitor.

Still another popular method of providing bias stabilization in transistor circuits is through a bias resistor placed in the emitter circuit. It is another form of self bias, and is generally used in conjunction with base bias. If the collector current flowing through the emitter resistor tends to increase, say as the result of an increase in temperature or a change in transistors, the voltage drop across R_e would increase. This would reduce the base-emitter bias voltage, resulting in a collector current decrease. The opposite would occur should the collector current decrease. The voltage drop across R_e reproduces a signal loss through degeneration. The emitter stabilizing resistor is often bypassed however, to avoid a-c signal degeneration. In this respect, the network operates similar to that of a bypassed cathode resistor in vacuum tube circuits.

Transistor Analysis Using Characteristic Curves

We are familiar with the construction of load lines in vacuum-tube circuitry and the analysis we can make from them. In transistors, a similar situation exists: we can construct a load line and use it for circuit analysis. Using a common-emitter circuit, we will calculate the current, voltage, and power gain from the output characteristic curves. These curves plot the collector current against the collector voltage for different values of base current. We will assume the following conditions: collector supply voltage (V_{CC}) is 10 volts; load resistor (R_L) is 3000 ohms; emitter-base input resistance (r_i) is 500 ohms; peak-to-peak input current is 20 microamperes. The operating point (X) is 25 μa base current .

COMMON-EMITTER AMPLIFIER
and its OUTPUT CHARACTERISTIC CURVES with LOAD LINE

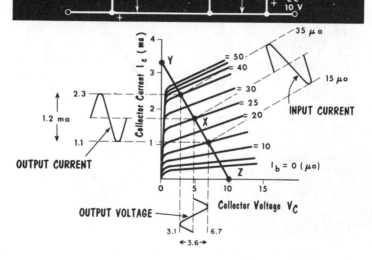

Our first step in constructing a loadline for R_L on the output characteristic curve is as follows: when I_C is zero, there is no voltage drop across R_L, and the collector voltage (V_C) is equal to the full 10 volts. We make this point Z on the collector voltage axis. When V_C is zero, the total V_{CC} is being dropped across R_L. Under this condition, I_C is equal to 10/3000 or 3.3 ma. We call this point Y on the I_C axis. Connecting points Y and Z establishes the load line. We now establish our desired operating point on the load line where I_b equals 25 μa. Vc is equal to 4.8 volts.

Transistor Analysis Using Characteristic Curves (Cont'd)

Since the input current is 20 μa peak-to-peak, it will deviate above 10 μa and below the operating point. Thus, the input current waveform will vary between 15 μa and 35 μa. Following this, we can construct the output current waveform by projecting across to the collector-current axis. The output current will swing between 2.3 ma and 1.1 ma. Projecting downward, we get the output voltage waveform and note that it swings between 3.1 and 6.7 volts.

Current Gain: In the common-emitter amplifier, this is the ratio of the change in collector current to a change in base current. From our construction curve, we find this is 1.2 ma/0.02 ma, or a current gain (amplification) of 60.

Voltage Gain: In the common-emitter amplifier, this is the ratio of the change in collector voltage to a change in base voltage. We can determine the change in base voltage since it is equal to a change in input current (20 μa) multiplied by the input impedance (500 ohms), or 0.01 volt. The voltage gain is thus equal to the ratio of 3.6 volts (ΔV_c)/0.01 volt (ΔV_b), or an amplification of 360.

Power Gain: This is equal to the voltage gain multiplied by the current gain (360 times 60), or 21,600. The power input is thus increased 21,600 times in going through the transistor. In terms of decibels, this is approximately 33 db.

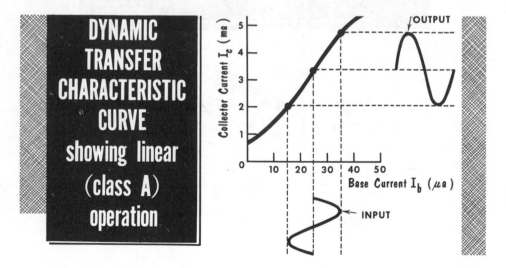

Another important curve for studying transistor characteristics is the dynamic transfer characteristic curve, in which collector current is plotted against base current for a given collector voltage. This curve is similar to the Ip-Eg vacuum tube curve. When the proper operating point is determined, and if the change of base current is within the linear portion of the dynamic transfer characteristic curve, the transistor operates linearly. This is called class-A operation, and the output signal will be an exact replica of the input signal.

Heat Sinks

Transistors are generally quite sensitive to increases in temperature. As the temperature of a semiconductor rises, an increasing number of electrons in the crystal are freed, producing an increase in collector current unrelated to the emitter or base currents. (This leakage current is normally quite small.) Leakage current flow can cause a further rise in the collector temperature, which releases still more electrons, again increasing the leakage current. Once started, this process (called thermal runaway) continues to repeat itself, regenerating until the transistor overheats and destroys itself. Because of this danger, the power-dissipation ratings of transistors are always given for a particular range of temperatures. Silicon transistors are usable at far higher temperatures than are their germanium equivalents.

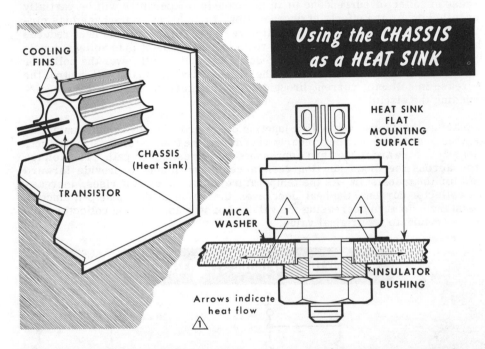

COOLING FINS

CHASSIS (Heat Sink)

TRANSISTOR

Using the CHASSIS as a HEAT SINK

HEAT SINK FLAT MOUNTING SURFACE

MICA WASHER

INSULATOR BUSHING

Arrows indicate heat flow

Power transistors are often mounted on a heat sink (usually a separate piece of metal or the chassis of an amplifier), to carry off the heat and to prevent the temperature of the transistor from rising above that of its surroundings. One power transistor, for example, requires a heat sink when dissipating 8.5 watts. If no heat sink is used, its maximum permissible dissipation in air is 1.5 watts. In many power transistors, the collector is connected internally to the case or shell. This acts as a considerable aid toward heat dissipation.

Heat sinks vary in shape and size, with power transistors usually being mounted on blocks of aluminum having many air-cooled fins. Another type of device, which is actually a finned heat radiator rather than a heat sink, fits tightly over the transistor case. The radiator contains many cooling fins that, effectively, increase the heat-dissipating area of the transistor case.

Thermistor and Diode Stabilization

Heat sinks reduce stabilization problems but they do not eliminate them. Another method is the use of temperature-sensitive control circuits. A temperature-sensitive device is a <u>thermistor</u> (thermal resistor), a device whose resistance changes with changes in temperature. Thermistors have a negative temperature coefficient; this means that as the temperature about the thermistor increases the thermistor resistance decreases.

A circuit using an unbypassed emitter resistor and a thermistor as part of the base bias circuit illustrates the use of two methods of thermal stabilization. The degenerative action of the unbypassed emitter signal in series-opposing that of the input signal has already been discussed. An undesired increase in collector current due to an increase in temperature will be partially cancelled due to the effects of degeneration. In addition, in the base bias circuit using the thermistor as the emitter resistor, the increased heat reduces the value of the thermistor, in turn reducing the value of bias voltage developed across the thermistor. The reduced base bias will lower the collector current flow, which in addition to the degenerative feedback will control the increase in collector current brought on by the increase in temperature surrounding the circuit.

Replacing the thermistor with a junction diode provides a p-n junction acting as a thermal device that more nearly resembles the transistor it is to protect. Diode D is forward biased and represents a low resistance path. The voltage drop across the diode junction resistance is sufficient to provide forward bias for the transistor. As the temperature about the circuit rises, increasing collector current, the heat decreases the value of the diode p-n junction resistance, in turn decreasing the value of base bias, causing collector current to return to its original value.

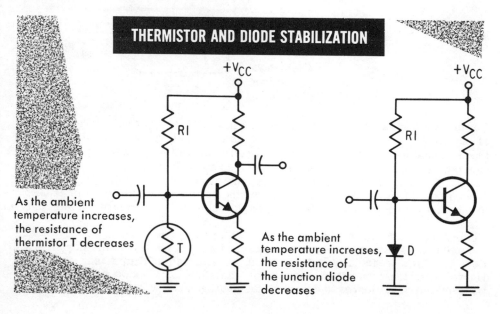

THERMISTOR AND DIODE STABILIZATION

As the ambient temperature increases, the resistance of thermistor T decreases

As the ambient temperature increases, the resistance of the junction diode decreases

Interstage Coupling

COUPLING CAPACITOR **Q2**
COUPLING RESISTOR
Q1
From preceding stage
R_{L1}
R_{L2}
To following stage
DECOUPLING NETWORK
POWER SUPPLY BYPASS CAPACITOR
V_{CC}

RESISTANCE-CAPACITANCE COUPLING

We have examined the basic types of transistor circuit arrangements. However, to obtain sufficient amplification or gain, more than one amplifier is required. As in tube circuits, two or more circuits can be cascaded for greater gain. The most popular methods of coupling transistor circuits are through R-C and transformer arrangements.

A typical two-stage R-C coupled common-emitter amplifier is shown. This type of coupling is desirable for the amplification of low-level or weak signals, where transformers would be more susceptible to hum pickup. Another feature is the fact that resistors and capacitors take up much less room than transformers. In our R-C coupled circuit, fixed bias is used together with emitter stabilization. The collectors of Q1 and Q2 are connected to the battery through load resistors RL1 and RL2. Since the coupling capacitor must be able to pass the lowest frequencies, and the input and output resistances are relatively low, the coupling capacitor must be fairly large in order to present a low reactance. Coupling capacitors are generally electrolyte types, and values of 10 μf are common. Being electrolytic, polarity always must be observed. Fortunately, the larger leakage currents of these capacitors are not as critical as in tube circuits.

To prevent undesired feedback, a decoupling network is often used. Decoupling can be achieved merely by placing a resistor in series with the base resistor and then bypassing the resistor. The time constant of this network must be long enough to fully bypass the lowest frequency. Since R must be kept small so as not to lower the battery voltage to the preceding stages, a very large decoupling capacitor of about 50 to 100 μf is frequently used. To prevent feedback because of the voltage drop across the internal resistance of the battery, a bypass capacitor is placed across it.

Another popular means of coupling transistor circuits is with transformers. Since the collector impedance of a transistor is high compared to the base input impedance, a transformer offers an excellent means of matching impedances and thus providing maximum power gain.

Interstage Coupling (Cont'd)

Although a stepdown transformer is used, this does not mean that there will be a signal loss. Since the transistor is a <u>current</u> device, the voltage stepdown transformer will actually provide a <u>current</u> gain for the signal. This action is similar to the output transformer in an amplifier that feeds a loudspeaker. A typical circuit might include a voltage divider for base bias and an emitter-stabilizing resistor, bypassed to prevent signal degeneration. The primary winding (including the a-c reflected load from the secondary) is the collector load impedance of Q1. The secondary winding introduces the a-c signal to the base and also acts as the base d-c return path.

Because there is no collector load resistor to dissipate power, the power efficiency of the transformer-coupled amplifier is high. However, the frequency response of this type of coupling is not as good as that of the R-C coupled stage. The primary winding shunt reactance at low frequencies reduces the low-frequency response. High-frequency response is reduced by collector capacitance and leakage reactance between primary and secondary.

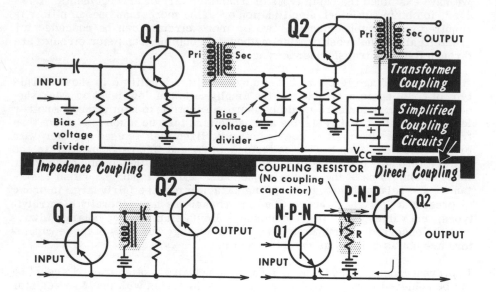

The impedance-coupled amplifier offers a compromise: an inductor replaces the collector load resistor. Thus, the d-c power loss is virtually eliminated; however, the low-frequency response is reduced by the shunt reactance of the inductor. The high-frequency response is reduced by the collector capacitance. Frequency response is better than that of the transformer arrangement, but not as good as that of the R-C coupled amplifier.

The direct-coupled amplifier is used for amplification of d-c and low-frequency signals. Its principal feature is that it retains the d-c component of a signal. Note that coupling capacitors are eliminated. Coupling resistor R acts as both the collector load resistor for Q1 and the bias resistor for Q2.

Interstage Coupling (Cont'd)

With direct coupling, the output of an amplifier stage is coupled <u>directly</u> to the input of the succeeding stage; no inductive or capacitive coupling is used. The main advantage of direct coupling is that the lowest frequencies are amplified equally as well as the highest frequencies.

A circuit using an NPN and a PNP transistor complementing each other illustrates a basic direct-coupled circuit. A disadvantage is the requirement of separate plus and minus V_{cc} sources.

In the preamplifier circuit shown the collector current flow of Q1 is carefully set so that with no signal applied the voltage drop is correct to properly bias the base of emitter-follower Q2. With Q2 an emitter-follower, its collector is directly connected to the $+V_{cc}$ source. The unbypassed emitter of Q1 has in parallel an R-C network to provide "equalization"; this refers to having the cartridge output signal levels at close to the original signal levels used during the recording. The design of the circuit is such that the input impedance of Q1 at 1 kHz is approximately 40K, at the low frequencies to 40 to 60 Hz the equalization circuit causes the input impedance to rise to approximately 650K. This accentuates the low audio frequencies which are usually lost or greatly reduced in R-C coupled amplifiers.

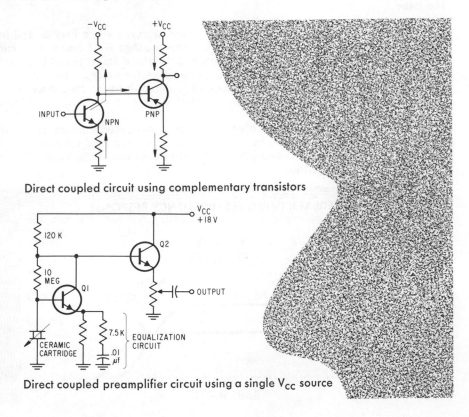

Direct coupled circuit using complementary transistors

Direct coupled preamplifier circuit using a single V_{cc} source

Negative Feedback

The reasons for the use of negative feedback to reduce distortion in audio amplifiers has been discussed for vacuum tube circuits in Volume 3; the same reasons prevail for transistor audio amplifiers. Negative feedback improves the frequency response (at the expense of amplifier gain), reduces distortion, and, most important, in transistor amplifiers feedback helps to reduce the differences in gain between transistors of the same type. Transistor uniformity, although constantly improving, still can vary widely between transistors of the same coded type, as well as between similar types. Frequency response of a single amplifier stage, or of a cascaded amplifier, appears as shown in the illustrated frequency response curves. The addition of negative feedback improves the overall frequency response curve at the expense of gain.

It was noted in the discussion of the common-emitter amplifier circuit, and in bias circuits, that the unbypassed emitter load resistor provides a negative feedback. Another method of providing feedback within a single stage combines both feedback and bias. In the circuit shown R_F is the feedback resistor, which in conjunction with R_1 also forms a voltage divider to set base bias. With a signal applied to the amplifier, a portion of the inverted output signal at the collector is fed back, $180°$ out-of-phase, to the input signal at the base.

In a cascaded amplifier the output of a second or third stage may be fed back to the input to provide degeneration. In the three-stage R-C coupled cascaded amplifier shown the output of the collector at the third stage is fed back to the input via an R-C coupling network as degenerative feedback. To provide degeneration the signal must be fed back $180°$ out-of-phase; therefore it must be provided by the output of the first or third stage.

In the two-stage cascade amplifier use of an unbypassed emitter resistor in the second stage provides feedback to Q2 via the unbypassed emitter resistor and at the same time the out-of-phase signal is coupled to the input circuit of Q1 for added degeneration.

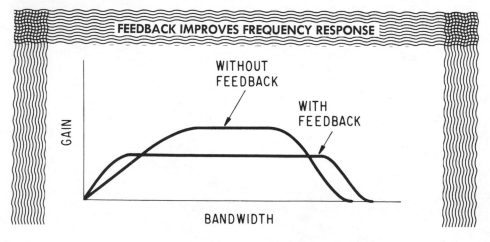

FEEDBACK IMPROVES FREQUENCY RESPONSE

WITHOUT FEEDBACK

WITH FEEDBACK

GAIN

BANDWIDTH

TYPICAL FEEDBACK CIRCUITS

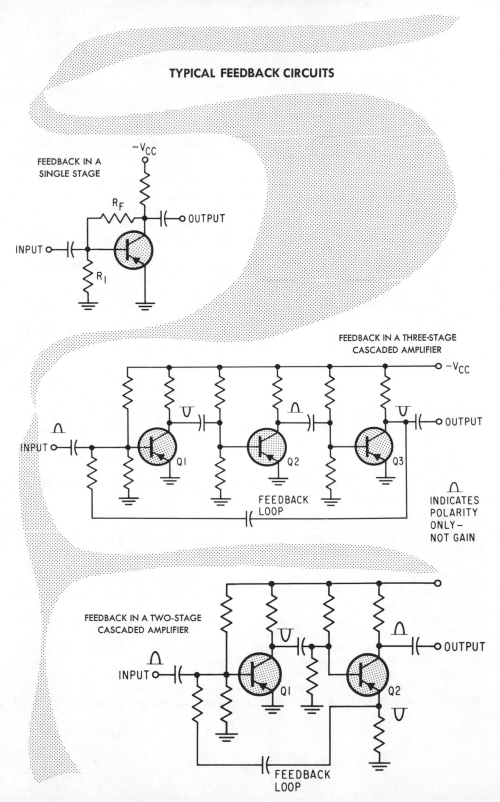

FEEDBACK IN A
SINGLE STAGE

FEEDBACK IN A THREE-STAGE
CASCADED AMPLIFIER

FEEDBACK
LOOP

INDICATES
POLARITY
ONLY–
NOT GAIN

FEEDBACK IN A TWO-STAGE
CASCADED AMPLIFIER

FEEDBACK
LOOP

Since there are both n-p-n and p-n-p transistors, a separate graphical symbol is used to represent both.

The transistor can be connected as a common-base, common-emitter, or common-collector amplifier.

The common-base and common-collector amplifiers do not provide a phase reversal of the input voltage signal.

The common-emitter amplifier provides a 180° phase reversal of the input voltage signal.

The circuit of the common-collector amplifier is similar to that of the cathode-follower vacuum tube amplifier, thus it is often referred to as an "emitter-follower."

The common-base and common-emitter circuits have low-to-medium input impedance and high output impedance; the common-collector circuit has high input impedance and low output impedance.

Reverse-bias collector current (I_{cbo}), also called saturation current, increases rapidly at high temperatures and causes increased emitter current.

D-c negative feedback can be used to minimize variations in emitter current caused by temperature changes.

The current, voltage, and power gain of a transistor amplifier can be calculated from the output static characteristic curves on which a load-line has been drawn.

The load-line indicates the way in which the collector supply voltage is divided between the load and the collector under various conditions of collector current.

The dynamic transfer characteristic curve may be used to determine the linearity and nonlinearity of the output signal to the input signal for a specific operating point and a specific load resistance.

Temperature stabilization can be achieved by using thermistors or diodes to vary the base bias as the temperature varies to keep the collector current stabilized.

Direct coupling operates without the use of frequency-restricting coupling circuits to amplify all frequencies without discrimination.

Negative feedback improves the frequency response of an amplifier at the expense of gain.

REVIEW QUESTIONS

1. Name the three basic-type circuits in which transistors can be connected.
2. A very low input impedance and a very high output impedance are characteristic of what type of amplifier configuration?
3. Give one important disadvantage of the common-base amplifier.
4. What is the phase relationship between the input and output voltages in the common-emitter amplifier?
5. In what type of circuit configuration are very high power gains possible?
6. Which circuit arrangement has a very high input resistance?
7. Which circuit arrangement always gives less than unity voltage gain?
8. What is meant by fixed base-current bias?
9. What is meant by self bias?
10. What is the function of the emitter-stabilizing resistor?
11. Give an advantage and a disadvantage of transformer coupling in transistor circuitry.

R-F Amplifier

A typical r-f amplifier circuit is shown. Using a high-gain ferrite-rod antenna in the antenna tuning circuit, the input signal is fed to the transistor base circuit through magnetic coupling. Through transistor action, the signal is amplified and then transformer-coupled to the input of the following mixer or converter stage. The stepdown r-f output transformer provides an impedance match between the high impedance of the r-f amplifier collector circuit and the low impedance of the following base or emitter circuit. Note that the collector is connected to some point on the output transformer primary. This is a design feature which permits finding the optimum point to present the best output load impedance for the collector.

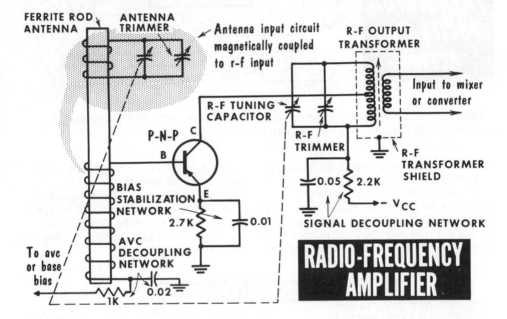

A tuner used for reception of FM stations is also illustrated. Note in the circuit the use of two tuned stages for improved selectivity. In this circuit the collector is returned to ground and $-V_{CC}$ is applied to the emitter.

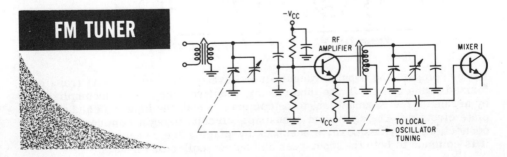

Transistor Oscillators

Oscillators have two principal sections; a frequency-determining section and an amplifying section. The frequency-determining (feedback) section usually consists of an L-C or R-C network. The amplifying section is usually a tube or transistor amplifier having sufficient gain to compensate for losses in the frequency-determining section. This arrangement provides us with positive feedback, which in turn produces oscillation. Thus, we can say that a tran-

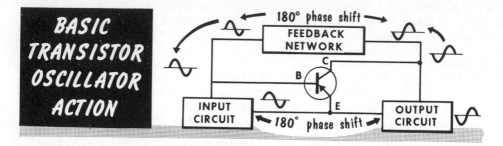

sistor oscillator is nothing more than an amplifier that has a portion of its output signal feeding back, in proper phase, to the input.

With the transistor acting as an amplifier, the input signal is amplified, with a portion of the output energy fed back to the input to supply the necessary input power to overcome circuit losses. When this is done, the transistor supplies its own input signal and oscillates at a frequency determined by the value of the feedback components. The transistor oscillates because any small current change in either the input or output circuit is transferred from one to the other through the transistor and feedback network.

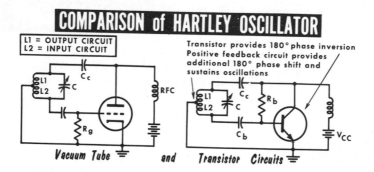

Let us compare the popular vacuum-tube Hartley oscillator with its transistorized counterpart. In the tube circuit, positive feedback is accomplished by arranging the resonant tank to be common to both the input grid and output plate circuits. The equivalent transistor circuit, using a common-emitter connection, provides positive feedback by placing the resonant tank so that it is common to both the input-base and output-collector circuits.

Colpitts and Phase-Shift Oscillators

The Colpitts oscillator is very similar to the Hartley oscillator. The major difference is that the feedback is taken from a capacitor voltage divider. In the circuit shown the tap between the two capacitors in the L-C circuit supplies the in-phase positive feedback to the emitter to sustain oscillations.

The phase shift oscillator utilizes the effect of an R-C phase shift to provide the 180 phase shift. A basic R-C network in which the capacitive reactance is large compared to the resistance (at the frequency of operation) will cause a phase shift due to the time taken to charge and discharge the capacitor. The most common method is to select values of R-C such that the phase shift of each R-C network is $60°$ at the selected frequency. The use of three R-C networks in series will then provide a $180°$ phase shift to provide positive feedback. In the circuit shown, the oscillations (oscillation will start due to random noise in the circuit) will provide a signal to the base that is then inverted $180°$ by the common-emitter amplifier. The inverted output at the collector of the amplifier is then fed back to the series of phase shifting R-C networks to provide another $180°$ of phase shift. With the phase-shifted signal returned to the base with $360°$ of inversion, the signal is in-phase for positive regeneration.

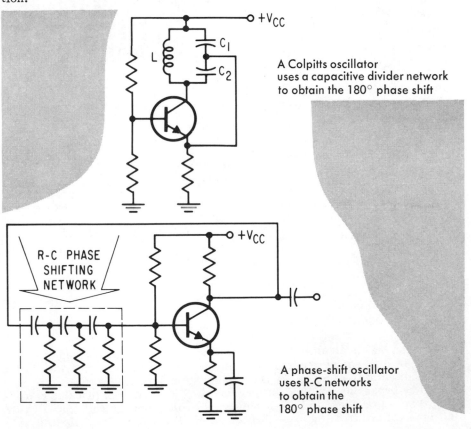

A Colpitts oscillator
uses a capacitive divider network
to obtain the 180° phase shift

R-C PHASE
SHIFTING
NETWORK

A phase-shift oscillator
uses R-C networks
to obtain the
180° phase shift

A Practical Local Oscillator

The common-emitter oscillator shown is a popular Hartley-oscillator type used in many transistor radios. It generates the unmodulated signal against which the modulated incoming signal is heterodyned in the mixer stage. The frequency-determining network consists of coil L and the oscillator tuning and trimmer capacitors. The oscillator coil, from point A to point C, resonates with the capacitors. Effectively, point C is at ground potential for a-c due to the low reactance of C1 at radio frequencies. This capacitor also prevents the collector bias from shorting to ground. The portion of the coil between points B and C forms the output circuit since it falls between the collector and the emitter. The portion of the coil between C and E forms the input circuit since it falls between the base and the emitter. The emitter, of course, is grounded through the stabilizing bypass capacitor and thus, is effectively at point C. Point D is used as a takeoff point for feeding the output of the oscillator to the mixer stage through capacitive coupling. Its exact position is determined by the circuit designer to provide the best combination of maximum signal output with optimum impedance match.

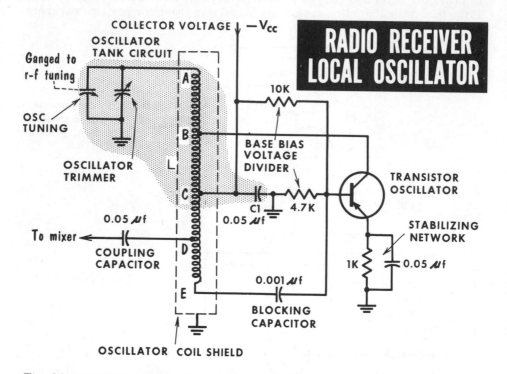

Fixed base bias is obtained by a voltage-divider network comprised of a 10K- and a 4.7K ohm resistor in parallel, with emitter stabilization provided by the 1000-ohm emitter resistor. Signal degeneration across this resistor is prevented by bypassing it with a .05-μf capacitor. The .001-μf blocking capacitor permits the signal to be fed to the base from point E, while blocking the full bias voltage from the base through points D and E in coil L.

The Mixer—Oscillator Circuit

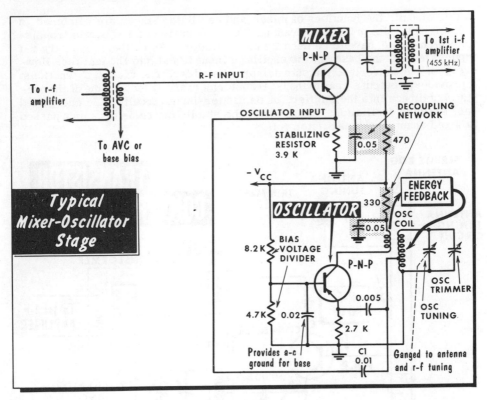

The function of the mixer stage is to heterodyne or mix the unmodulated local oscillator output signal with the incoming signal from the r-f amplifier. Many frequencies will be present in the output circuit as the result of the hetero-dyning process, and it remains for the i-f amplifier to select the desired frequency. The mixer stage, basically an amplifier having a tuned output circuit, is biased on a nonlinear portion of its characteristic curve (linear amplification does not produce heterodyning). By having a separate oscillator stage, the oscillator is usually apt to be more stable and unaffected by changes in other circuits.

In the typical circuit, the output from the r-f amplifier (or directly from the antenna) is fed to the base of the mixer circuit. The output from the oscillator is fed to the emitter of the mixer circuit. Thus, effectively, the two signals are mixed in series. Coupling capacitor C1 permits the passing of oscillator energy while blocking the low resistance of the oscillator coil from shunting the stabilizing resistor. The oscillator is a common tickler-feedback type, with energy from the collector circuit inductively coupled back to the emitter. The base of the oscillator is effectively at a-c ground. The usual bypass capacitor across the stabilizing resistor is eliminated to prevent the oscillator signal from being shunted to ground.

The Converter

In this circuit, the functions of mixer and oscillator stages are combined in a single unit. Its principal advantage is in the saving of a separate transistor which would have to be used as an oscillator. As in the mixer, the r-f input is fed into the base and the oscillator input is fed into the emitter. However, in the converter, the output from the collector serves two functions: it provides an output signal to the i-f transformer; and it feeds some of the output signal back into the emitter, or oscillator-input circuit. The energy fed back sustains oscillations. Once again, the stabilizing resistor is unbypassed to avoid shunting the oscillator coil.

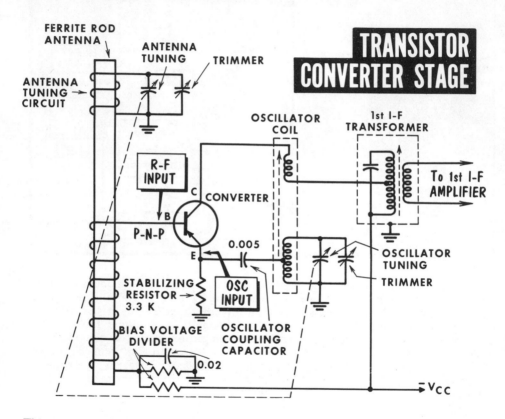

This circuit is popular since most radios do not require a stage of r-f amplification. Thus, this circuit and variations of it are found in virtually all "economy-type" receivers. There can be many variations of the oscillator circuit, but almost all operate on the principle that there is positive feedback from the collector-output circuit to the emitter-input circuit. Biasing is arranged so that the base-emitter characteristic is nonlinear, resulting in heterodyning action. The 1st i-f transformer is tuned to the intermediate frequency, usually 455 kHz. This transformer is a stepdown unit providing both a current gain and an impedance match between the convertor collector circuit and the first i-f amplifier input circuit.

The I-F Amplifier

As we have learned from our study of vacuum-tube circuits, the i-f amplifier is nothing more than a fixed-tuned r-f amplifier. Its function is to select one of the many signal frequencies present in the output circuit of the converter stage, and to produce high amplification of this signal. The selected signal represents the difference frequency between the incoming r-f signal and the oscillator output. Being fixed-tuned, the i-f amplifier need not amplify a wide range of frequencies, but merely be tuned to one--the intermediate frequency (usually 455 kHz). As such, all tuned circuits in the i-f amplifier (usually permeability tuned) are adjusted to peak at one single frequency, and to provide maximum gain at that frequency. The Q of the coils used in these stages can be quite high, since the band of frequencies passed need be only 10 kHz (±5 kHz from the i-f).

Because of the high gain of the i-f amplifier, the stages often tend to become unstable and oscillate, causing howling in the audio output. To avoid this, many i-f amplifiers use negative feedback; that is, they return a portion of the output (out of phase) to the input. In the circuit shown, negative feedback, or collector neutralization, is accomplished through a capacitor connected between collector and base in each stage. This cancels any positive feedback voltages developed. Two stages of i-f amplification are typical for most transistor receivers. In some circuits, the avc voltage is applied to both stages; in others it is applied to only one. Note that in this circuit, the emitters are returned to a positive voltage and the collectors to ground. This indicates that the negative side of the battery is grounded.

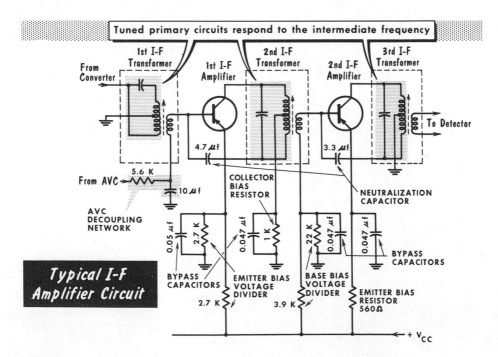

I-F Amplifer with Overload Diode

The two-stage i-f amplifier shown here incorporates an interesting feature found in many transistor radios--the use of an <u>overload diode</u>. Indicated in the diagram as D1, it is connected from the collector of the converter to a tap on the primary winding of the first i-f amplifier output transformer. In this position, the diode is effectively connected across a portion of the primary of the converter output transformer, and acts as a variable r-f load on the transformer. D1 is biased so that it will not conduct on weak signals. As we will see, this diode prevents overloading on strong signals.

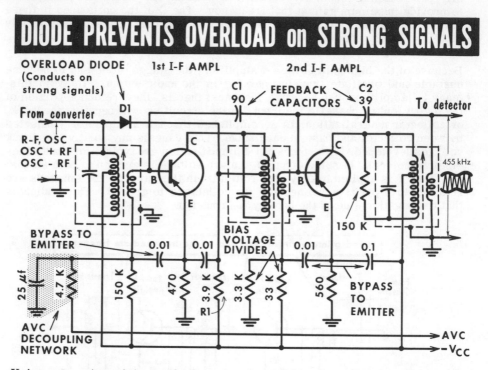

DIODE PREVENTS OVERLOAD on STRONG SIGNALS

Using p-n-p transistors, the base is always biased <u>negatively</u> with respect to the emitter. When a strong signal is received, the <u>positive</u> avc voltage applied to the base of the first i-f amplifier is increased, thus decreasing the forward bias on that transistor. The decrease in forward bias results in a decrease in collector current, which in turn lowers the voltage drop across R1 and increases the collector voltage on the first i-f amplifier. This produces a reduction in bias across D1. On strong signals, the i-f amplifier collector voltage will approach the collector voltage of the converter stage. When this occurs, the overload diode bias is cancelled and D1 begins to conduct and load down the primary of the first i-f transformer. This damping, or loading down, lowers the gain of the circuit and compensates for the very strong signal. Also note in this circuit the feedback capacitors. This neutralization or feedback is not required for all circuits and depends upon transistor and circuit characteristics.

The Detector-- AVC

DIODE DETECTOR and AVC CIRCUIT

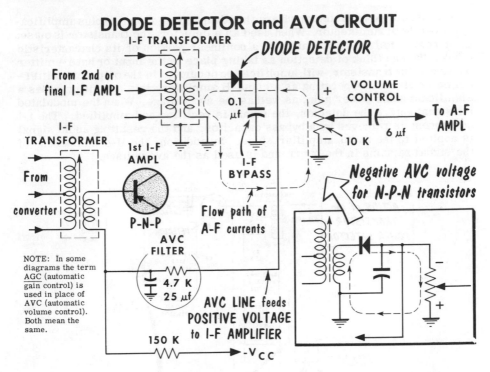

NOTE: In some diagrams the term AGC (automatic gain control) is used in place of AVC (automatic volume control). Both mean the same.

The purpose of the detector is to obtain from the modulated i-f an undistorted copy of the modulation waveform, which represents the signal intelligence. Frequently, the detector also has to produce an output signal proportional to the carrier amplitude which can be used for automatic volume control. Both diodes and transistors can be used for detection and avc. Avc is used to minimize the effects of fading, and to obtain approximately the same output volume from signals of different strengths. This control is obtained by feeding back to one or more i-f stages a voltage proportional to the carrier strength at the detector load. The control voltage has the effect of reducing the gain of the controlled stages. The most popular method of doing this is by reducing the emitter current in these stages.

In our typical circuit, the i-f signal is rectified by the diode detector, the i-f component being bypassed by the 0.1-μf capacitor. The audio component is then developed across the 10,000-ohm volume-control potentiometer. When p-n-p transistors are used as i-f amplifiers, the base is biased negatively with respect to the emitter. Thus, since the function of the avc voltage is to reduce the gain of the amplifier, the avc voltage fed to the base must be positive so as to reduce the forward bias on the transistor. In n-p-n transistors, the avc voltage must be negative. This can be done merely by reversing the polarity of the diode connection. The avc filter removes signal variations and tends to keep the avc voltage steady.

The Power Detector

The transistor power detector offers the advantage of detection plus <u>amplification</u> over the diode detector. When used as a detector, the transistor is biased at or near cutoff, thus operating in a nonlinear portion of its characteristic curve. We can think of detection as taking place in the input or base-emitter portion of the transistor, with amplification occurring in the output or emitter-collector circuit. Operating as a class-B amplifier, this circuit provides a significant audio power gain, as well as the avc voltage. When the modulated i-f signal enters the detector, the signal is rectified and amplified. The i-f component is removed by a bypass capacitor, and the resulting audio signal is applied to the audio amplifier stage. In addition, the d-c component of the signal remains in the output and is used as the avc voltage.

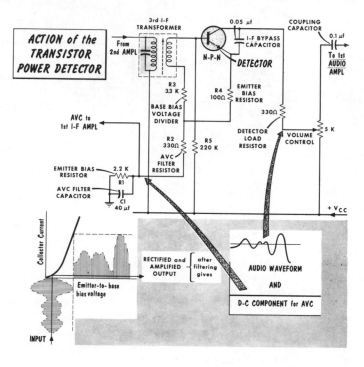

In the circuit shown, cutoff bias is applied to the base of the detector through R1, R2, and voltage divider R3-R5. When the incoming signal strength increases, the signal fed to the detector base-emitter circuit increases. This produces an increase in detector current during the conducting half cycle, and current will flow up from ground through R1, R2, R3, and then from emitter to collector in the detector. This increase in current through R1 produces an increase in <u>positive</u> bias which can be applied to the emitter of an n-p-n transistor or the base of a p-n-p transistor, reducing the forward bias and lowering the gain of the i-f stage. C1 and R2 act to present the proper time constant for avc filter action. The opposite will occur when a weaker signal arrives at the detector.

The Audio Amplifier

THE AUDIO DRIVER STAGE

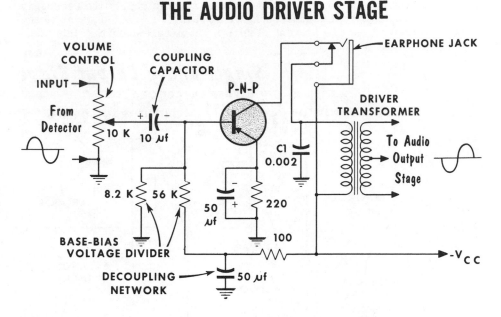

The output of the detector stage consists of an extremely weak audio signal that represents the modulation of the incoming r-f signal. What is required at this point is a stage of audio amplification to build up this small audio signal to a point where it can drive the power output stage. The power output stage, single-ended or push-pull, requires far more driving power than can be delivered by the detector stage directly, except in some cases where a power detector is used. In most cases, one stage of audio amplification is required; however, a two-stage resistance-coupled amplifier is not unusual.

In the typical audio driver stage shown, the input signal is taken from the sliding contact of the volume control and fed to the base of the audio amplifier through a 10-μf coupling capacitor. This capacitor isolates or blocks the d-c base-bias voltage of the audio stage from the detector and volume-control circuitry. The base is biased through conventional voltage-divider bias, and the emitter contains a 50-μf capacitor bypassing a 220-ohm stabilizing resistor. Transformer coupling is used to match the high collector output-circuit impedance of the audio amplifier to the low input impedance of the output stage.

An interesting feature found on many sets is the earphone jack in the audio stage. While the output of this stage is insufficient to drive a loudspeaker, it can actuate earphones. When the phones are not inserted, the jack is short-circuited and there is normal circuit action. When the phones are inserted, the primary of the driver transformer is open and the phone impedance forms the collector load of the audio stage. C1 is used to stabilize the collector circuit and prevent oscillation; it also acts to improve tone response.

Single-Ended Audio Output Stage

Most transistor radios use a push-pull output stage to supply the necessary power needed to drive the loudspeaker. However, it is not unusual to see a single-ended stage. To prevent distortion, it is necessary that this stage

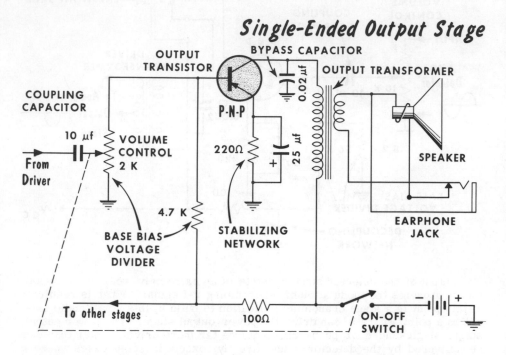

Single-Ended Output Stage

be driven as a class-A amplifier, where collector current flows throughout the entire input cycle. While termed a "power" amplifier, the typical portable receiver using a single transistor in its output stage is only capable of about 100 milliwatts output--small by vacuum tube standards but adequate for small 3-inch speakers. Where an appreciable amount of heat is to be dissipated by a power transistor, a heat sink is required. This is a metal conductor that will remove the excess heat. Often, the metal chassis is used as a heat sink. Without an adequate heat sink, thermal runaway may develop. This condition occurs when excessive heat causes increased collector current, which in turn causes still more heat and still more current.

In the typical single-transistor audio output stage, the input can come directly from the detector, or from a previous audio amplifier. The signal is coupled to the base of the output stage, with the volume control forming part of the base-bias voltage divider. Here again, the emitter stabilizing resistor is bypassed to avoid signal degeneration. A conventional output transformer matches the collector impedance to the speaker coil impedance. The transformer primary is bypassed to ground to avoid noise voltages and prevent high-frequency distortion. The jack permits listening with earphones and muting the speaker.

The Push-Pull (Double-Ended) Amplifier

DOUBLED-ENDED (PUSH-PULL) OUTPUT STAGE

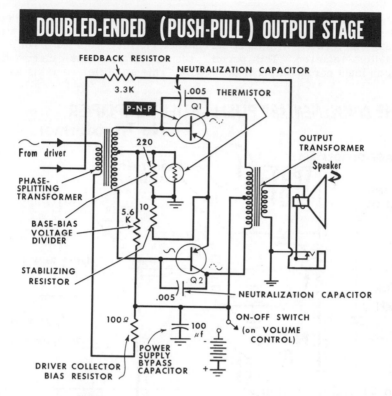

This amplifier consists of two transistor circuits operating 180° out of phase, but with their outputs combined. It can be operated either Class A or Class B, with the latter offering the advantage of efficiency in that collector current flows in each transistor only during alternate half cycles. A feature of push-pull circuitry is the elimination of even-order harmonics and the d-c component in the load. In Class-B operation, the output power can be as much as four times the collector dissipation of each transistor.

In our circuit, each transistor is biased near cutoff (Class B), and a split-secondary input transformer feeds out-of-phase signals to Q1 and Q2. Amplification takes place during the half cycle that each transistor conducts, and the outputs are combined in the secondary of the output transformer. Note the thermistor connected in parallel with one of the base-bias resistors. The thermistor resistance decreases as the temperature of this resistor increases, thus lowering the resistance of the parallel combination (which in turn reduces the base-bias voltage). The reduction in base bias limits current flow in the collector circuits, keeping the power dissipation of the transistors within the proper limits for the temperature at which they are operating. The feedback loop from one end of the output transformer secondary to the driver-stage emitter provides some inverse feedback which stabilizes overall operation of these circuits.

Complementary Symmetry Push-Pull Circuit

This unusual circuit has no parallel in vacuum-tube circuitry. It depends upon the complementary nature of transistors. That is, an n-p-n and a p-n-p transistor can be made with identical characteristics except for the difference in polarities required. This permits the use of a push-pull circuit having neither an input nor an output transformer. This amplifier contains an n-p-n

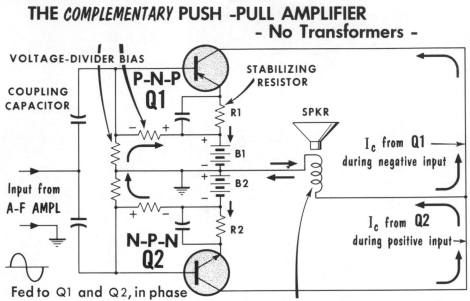

THE *COMPLEMENTARY* PUSH -PULL AMPLIFIER
- No Transformers -

VOLTAGE-DIVIDER BIAS

P-N-P Q1   STABILIZING RESISTOR

COUPLING CAPACITOR   R1   SPKR

B1 I_c from **Q1** during negative input

Input from A-F AMPL   B2

I_c from **Q2** during positive input

N-P-N Q2   R2

Fed to Q1 and Q2, in phase

Output signal appears in form of I_c fluctuations in Q1 and Q2

and a p-n-p transistor, each connected in a common-emitter stage. Conventional voltage-divider base bias is used for each transistor, with bypassed stabilizing resistors in each emitter circuit. A single input feeds the same signal to the base of each transistor, and a high-impedance loudspeaker voice coil acts as the output load.

We can analyze this circuit by considering it as a Class-A amplifier. Thus, with zero-signal input, both transistors are forward biased and collector current flows. Under this condition, there is perfect balance, and current flows from B2 (-) through R2, Q1, Q2, R1, and back to B1 (+). Since this now acts as a balanced Wheatstone bridge, no current will flow through the load or speaker coil. With a positive-going signal applied to both transistors, forward bias on the p-n-p transistor decreases, while forward bias on the n-p-n unit increases. Since the forward bias on Q2 is now greater than that on Q1, collector current from Q2 will flow through the speaker coil. When a negative-going signal is applied, the opposite occurs, and Q1 collector current flows through the coil. Thus, the current fluctuations through the coil are determined by the amplitude and frequency of the incoming signal, and audio signals are reproduced by the loudspeaker.

Power Sources

Most transistor radios are battery operated. A single battery may be used or several cells may be connected in series. The most common voltages used are 6 and 9 volts, while voltages of 3, 4, 5, and 9 are not unusual. The most popular battery is the special 9 volt battery referred to as a "transistor" battery. A popular power source is the 1.5 volt zinc-carbon cell, A, C, and D size. Another popular source of power is the mercury cell, having a voltage rating of about 1.34 volts. This also can be connected in series for higher voltage ratings. Because of the low current drain from a typical portable transistor radio—often less than 10 mA—battery life often may be as long as 400 hours using large cells such as the D-size zinc-carbon unit. Another popular cell is the 1.25 volt nickel-cadmium type. This is a secondary cell and can be recharged often by plugging it into a special recharging circuit operated from a 117 volt a-c supply.

It is most important that battery polarity always be observed. Whether n-p-n or p-n-p transistors are used, a battery can have either its positive or negative terminal grounded or common. Failure to observe this may permanently damage the transistors. For instance, a p-n-p transistor requires that the collector be negative with respect to the emitter. If the positive battery terminal is grounded, the emitter will be connected directly or through a network to ground. If the negative terminal is grounded, the collector circuit will connect to ground. Most transistor receivers have an on-off switch ganged to the volume control, and connected so as to open and close the battery circuit.

A "battery-eliminator" is often used where fixed operation of the transistor receiver is required. Using a 110 to 6.3 volt center-tapped filament transformer the circuit shown will supply a full-wave rectified voltage of the desired value to a plug-in jack mounted on the receiver. With the battery eliminator jack plugged in, the receiver battery is disconnected.

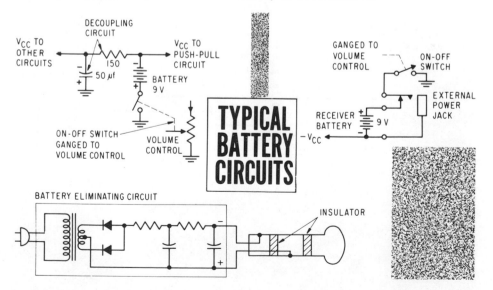

DECOUPLING CIRCUIT

V_{CC} TO OTHER CIRCUITS

V_{CC} TO PUSH-PULL CIRCUIT

150 50 μf BATTERY 9 V

ON-OFF SWITCH GANGED TO VOLUME CONTROL

VOLUME CONTROL

TYPICAL BATTERY CIRCUITS

GANGED TO VOLUME CONTROL ON-OFF SWITCH

RECEIVER BATTERY 9 V $-V_{CC}$

EXTERNAL POWER JACK

BATTERY ELIMINATING CIRCUIT

INSULATOR

AM-FM Superheterodyne Receiver

The pocket-sized AM-FM receiver circuit illustrated has some interesting features.   The FM antenna is an open loop of wire projecting from an upper corner of the receiver and acts as both antenna and carrying handle.   Transistor Q1 is the FM r-f amplifier, and for the high radio frequencies of FM (88-108 MHz) a common-base amplifier is used. Transistor Q2 is the FM converter, also a common-base configuration.   Positive feedback for Q2 is by capacitor coupling from the FM oscillator tuning circuit to the emitter input circuit. The variable capacitor, trimmer capacitor, and inductor in the collector circuit of Q2 are tuned to the local oscillator frequency, 10.7 MHz above the incoming radio frequency. The primary of the FM i-f transformer, T1, is tuned (with stray and circuit capacitances) to the difference frequency of 10.7 MHz.

Transistor Q3 is an i-f amplifier for FM and a converter for AM.   The 10.7 MHz output is coupled via T1 to the base of Q3. With the switch in the FM position the collector of Q3 has as a collector load the primary of the 10.7 MHz i-f transformer T3, and both the coupling winding of T4 and primary winding of T5; T5 being the 455 kHz AM i-f transformer.

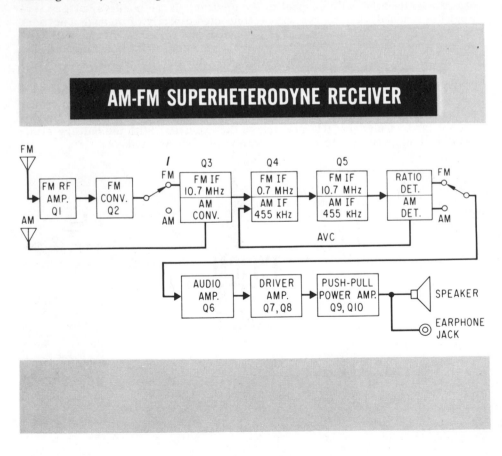

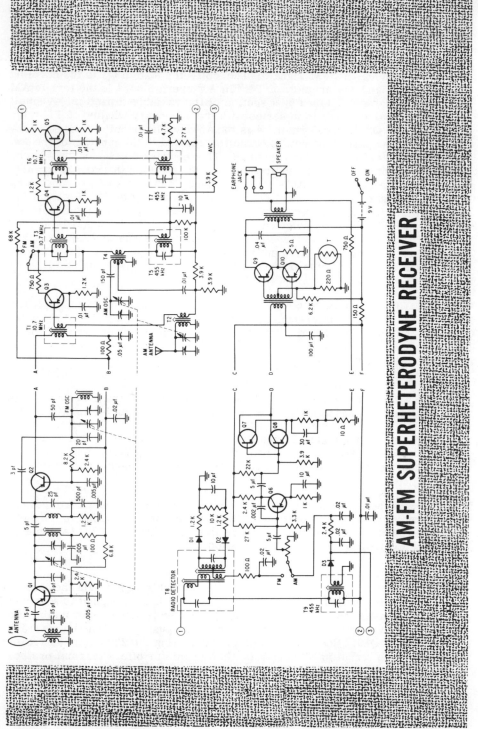

AM-FM SUPERHETERODYNE RECEIVER

AM-FM Superheterodyne Receiver (Cont'd)

With the AM-FM switch in AM, the primary of FM i-f transformer T3 is shorted, removing the FM signal. The collector load of Q3 is the coupling winding of T5 and the primary of T4. In AM operation T2 is the ferrite AM broadcast antenna coil, tuned by a section of the variable capacitor. A small secondary winding of T2 is in series with the secondary winding of T1 in the base circuit of Q3. Transformer T4 is the AM oscillator coil. A tapped portion on the lower part of the coil is capacitor coupled to the junction of the two 3.9K base bias voltage divider resistors. The feedback portion of the AM oscillator coil voltage is developed across the 3.9K resistor that goes to ground. This places a portion of the AM oscillator signal (455 kHz above the incoming AM signal) in series with the AM signal in the secondary of T2. Both the AM oscillator and r-f signals are applied to the base of Q3, with the difference frequency of 455 kHz being selected by the primary of the input AM i-f transformer T5. Since the collector circuit has the AM oscillator frequency, the AM radio frequency, and the sum and difference frequencies, there will be feedback from the oscillator frequency via the secondary winding of T4.

With the switch in the AM position the 455 kHz i-f output of T5 is amplified by Q4, coupled via the interstage AM i-f transformer T7, amplified again by Q5, with the output of T9, the output AM i-f transformer, detected by diode D3. The detected output of D3 is applied through the AM-FM switch to the audio amplifier section.

With the switch in the FM position, T1 is the FM input i-f transformer, Q3 is an i-f amplifier, T3 an interstage FM i-f transformer, Q4 an i-f amplifier, T6 an interstage FM i-f transformer, Q5 an i-f amplifier with T8 a ratio detector transformer, and diodes D1 and D2 the FM detector diodes. (For an explanation of the operation of the ratio detector refer to Volume 4.) The detected output is applied through the AM-FM switch to the audio amplifier.

Note that with the switch in the AM position there is AVC feedback from diode D3, filtered by a 3.9K resistor and a 10 μ f capacitor and applied to the base circuit to regulate the gain of Q4. There is no circuitry for AFC (Automatic Frequency Control) or AVC in FM operation.

Audio circuitry is conventional. The selected output is applied via the volume control to audio amplifier Q6. The output of Q6 is r-c coupled to a pair of paralleled transistors, Q7-Q8, with the paralleled transistors providing the driving power via an interstage transformer to the push-pull audio output transistors Q9 and Q10. Note the thermistor paralleling the 220 ohm bias resistor in the base circuit of Q9 and Q10, the thermistor offers thermal protection. Degenerative feedback to improve audio response is supplied by the feedback capacitor between the collector and base of Q6. Paralleled transistors Q7-Q8 have two resistors in the emitter circuit, one of which is bypassed. The unbypassed resistor receives feedback from the secondary winding of the output transformer.

Neutralizing circuits are used to eliminate the possibility of oscillations in high-frequency tuned amplifiers.

Small compact ferrite rods used as the core for AM antenna coils permits design of ultra-compact transistorized AM receivers.

Oscillators usually have two principal sections; a frequency-determining section and an amplifying section.

By utilizing the time required to charge and discharge a capacitor a phase-shift oscillator utilizes RC networks to create the 180° phase-shift required for oscillation.

The mixer stage is fundamentally an amplifier having a tuned output circuit, and biased on a nonlinear portion of its characteristic curve.

A converter stage utilizes a single transistor to act as both oscillator and mixer combined.

To avoid oscillation, many transistor i-f stages use negative feedback or neutralization.

The transistor power detector offers the advantage of detection plus amplification.

Transistor receivers utilize special shorting jacks in the audio output circuits to allow listening to an external earphone while muting the speaker.

Where an appreciable amount of heat is to be dissipated by a power transistor, a heat sink is required.

In class-B operation, the output power can be as much as four times the collector dissipation of each transistor.

Complementary symmetry output stages take advantage of the complementary behavior of both NPN and PNP transistors to provide an output stage requiring no input or output transformers.

Popular cells used in transistor radios include the zinc-carbon, mercury, and the nickel-cadmium type.

Battery eliminator circuits allow the use of a-c line operation by plugging the rectified d-c voltage into a special jack to eliminate the need for a battery.

AM-FM superheterodyne receivers utilize the same IF amplifier and audio amplifier circuits for both AM and FM while using separate RF tuned circuits and detector circuits for AM or FM.

REVIEW QUESTIONS

1. What is the function of the transistor r-f amplifier stage?
2. Explain the basic operation of the transistor oscillator.
3. What is the function of the mixer stage?
4. Basically, how does the converter differ from the mixer?
5. Why do i-f amplifier stages tend to be unstable?
6. Explain the operation of the overload diode.
7. What is the advantage of the power detector over the diode detector?
8. Explain the operation of the automatic volume control circuit.
9. What is meant by thermal runaway and how is it controlled?
10. Explain the operation of the complementary symmetry push-pull circuit.
11. Why is it important to observe battery polarity in a transistorized circuit?

Zener (Avalanche) Diode

The zener, or _avalanche_, diode is similar in appearance to the silicon or germanium rectifier diode.   Most often, it is made of silicon and has a very high back resistance.   When a reverse voltage is applied to this diode, virtually no reverse current flows. However, at a certain reverse voltage point, the zener diode breaks down completely, and the back resistance drops to a very low value.   When this occurs, the reverse current increases very rapidly ("avalanche" effect).   The effect of a rapid increase in current, together with a rapid decrease in resistance, produces an almost constant voltage drop across the diode. Thus, when biased in a reverse direction, zener diodes can be used as voltage regulators.

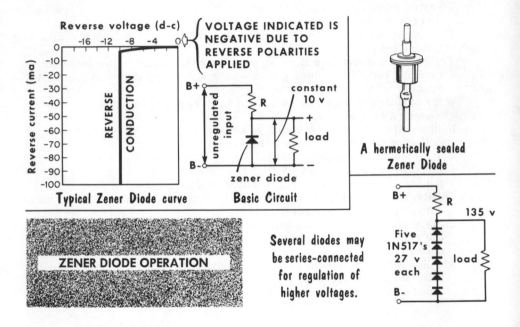

Typical Zener Diode curve   Basic Circuit

A hermetically sealed Zener Diode

ZENER DIODE OPERATION

Several diodes may be series-connected for regulation of higher voltages.

Beyond the breakdown or zener voltage, the zener diode exhibits the characteristics of a gas-voltage regulator, and can be considered an equivalent. To use the zener diode in a VR circuit, positive voltage is applied (through a series resistor) to the cathode. (This is opposite to the normal application of voltage to a diode rectifier.) The current flowing through R equals the sum of the current through the diode plus the load current. When the B voltage drops below normal, the voltage across the diode drops.   This increases the diode resistance, and less current flows through the diode and R. The reduced current lowers the IR drop across R, dividing the output voltage so that 10 volts, for example, is again across the diode.   When the output voltage rises above normal, the diode permits more current flow, its resistance decreases, and the current through R increases.   This increases the IR drop across R, and the 10-volt output across the diode is maintained.   The load, connected in parallel across the diode, has a fixed 10-volt drop maintained across it.

Tunnel Diode

The tunnel diode is a special diode named after the apparent "tunneling" effect that takes place in the diode junction. The N and P materials are specially doped to create a very narrow junction.

The tunnel diode curve is unusual in that current starts to rise in a normal manner, then as the voltage is increased the current drops, and then reverses and starts to rise again. The area in which current decreases as voltage increases is called the "negative conductance region." The current level where the negative conductance starts is called the peak current (I_p), and the voltage at this point is called the peak voltage (Ep). The current at the point where the negative conductance ceases is called the valley current (I_V) and the voltage at this point is called the valley voltage (E_V).

With the small value (tenths of a millivolt) of forward bias applied the barrier junction is not overcome, but the very thin junction area permits the current carriers to "tunnel" through the junction to provide the forward current that peaks at I_p. An additional increase in bias reduces the number of available electrons in the n-type material, reducing the flow of "tunneling" carriers; a continued increase in bias reduces the flow until the forward current is minimum at I_V. Further increasing the bias finds the diode acting as a conventional junction diode, the barrier is overcome and forward current flows as in a conventional diode.

Tunnel diodes are used in very high frequency circuits as oscillators and amplifiers. Their limited operating voltages restrict their use to very low power circuits.

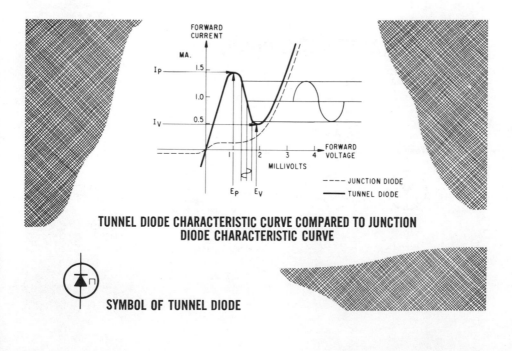

TUNNEL DIODE CHARACTERISTIC CURVE COMPARED TO JUNCTION DIODE CHARACTERISTIC CURVE

SYMBOL OF TUNNEL DIODE

Silicon Controlled Rectifier (SCR)

The silicon controlled rectifier (SCR) is a four-layer (PNPN) silicon diode
with a "gate" lead to one of the layers. The SCR is essentially a controlled
rectifier in that it will not pass current until "turned-on" by the gate signal,
and when turned-on the SCR provides a low-resistance current path.

With the anode voltage positive in respect to the cathode, application of a
small value of positive voltage to the gate will forward-bias the SCR to have
it turn on. Once the SCR is conducting, the gate voltage can be removed and
the SCR will continue to conduct (thus a voltage pulse can be used to trigger
the SCR) until the anode voltage is removed, or made very low.

The SCR characteristic curve is unusual. With the gate voltage applied the
initial forward current flow is very small, then suddenly as the current
reaches a certain point "avalanche" occurs and SCR current leaps to the
amount limited by the value of the load resistance in series with the SCR. The
value of anode-to-cathode voltage at the point where the current avalanches is
called the "forward breakover voltage," which refers to the fact that at avalanche the voltage across the SCR breaks down (breaks over) to a low value
and remains at the low value during operation. The result is the unusual
curve, where current flow is small up to the breakover point, then the current increases but the current curve snaps back to the lower operating voltage
across the SCR.

Note in the characteristic curve that an increased value of gate signal voltage
will cause an earlier breakover point.

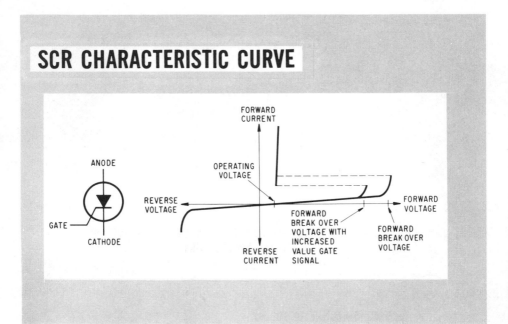

Unijunction Transistor (UJT)

The unijunction transistor (UJT) is not really a transistor, it is a three-ter-minal device containing a single junction. It is the combination of a single junction with three leads that gives it the name unijunction transistor. The UJT is constructed of a bar of n-type silicon with connections at each end, labelled base 1 (B1) and base 2 (B2), and a diode formed by a junction of p-type material at approximately the center of the n-type bar. The n-type bar is lightly doped and has a resistance of about eight thousand ohms. Applica-tion of voltage between B1 and B2 (positive to B2) will make the n-type bar a voltage divider, with the voltage at the junction about half the applied value. For example, with 8 volts applied between B1 and B2, the voltage on the bar opposite the junction will be about 4 volts.

With the 8 volts applied between B1 and B2, and no voltage applied to the emit-ter, the junction diode is reverse-biased and, aside from the very small emit-ter-base reverse current, the only current flowing is that in the relatively high resistance circuit of B1-B2. Application of just over +4 volts (or higher) to the emitter will forward-bias the junction diode. The forward biased emit-ter-to-B1 circuit will result in a heavy emitter-to-B1 current flow. The in-creased current flow through the portion of the bar between the diode junction and the B1 connection can be viewed as a reduced emitter-to-B1 resistance. This reduced resistance also increases the B1-to-B2 current flow. The in-creased value of current flow continues as long as the diode junction remains forward-biased.

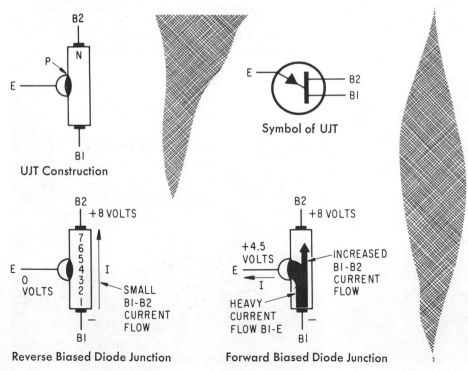

UJT Construction

Symbol of UJT

Reverse Biased Diode Junction

Forward Biased Diode Junction

Unijunction Transistor (UJT) (Cont'd)

Because of the diode conducting action at a fixed voltage the UJT is often used as a "trigger" to apply current to other circuits or devices. In addition to triggering, the UJT is often used as a relaxation oscillator. In the circuit shown the values of R1 and C1 determine the time required for C1 to charge through R1 to the value that will forward-bias the diode junction. Varying R1 will, to a slight degree, vary the frequency of the oscillator. When the junction is forward-biased, capacitor C1 discharges through the low resistance offered by the emitter-to-B1 circuit. When the capacitor voltage falls below the forward-bias value the diode becomes reverse-biased and the UJT ceases conduction. With the diode reverse-biased, the capacitor starts to charge again and the cycle repeats.

The voltage waveform at the emitter (across the capacitor) will be that of the capacitor charging and discharging to form a "sawtooth" waveform. With equal value resistors in the B1 and B2 circuits the resulting base current will provide output voltage pulses across R2 and R3 that have B1 and B2 provide output pulses of equal value but 180° out of phase with each other.

The basic triggering circuit shown has a UJT controlling a high current-handling transistor to operate an electromechanical relay. A low current positive voltage level input r-c coupled to the emitter of the UJT will provide a medium-current output from B2 of the UJT. The output of B2 is direct-coupled to the base of a power transistor. With the UJT conducting, the output level at B2 forward-biases the transistor, causing it to conduct. With the transistor conducting, the current flow through the coil winding activates the relay. Lowering the level of the input to the UJT causes it to cease conducting; in turn the transistor ceases to conduct and the relay opens.

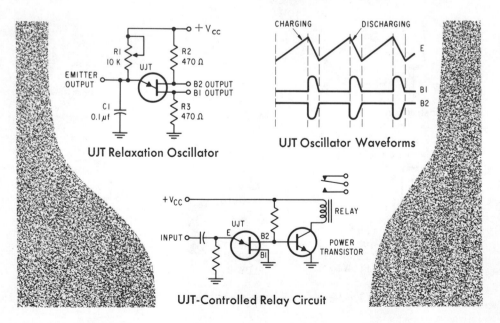

UJT Relaxation Oscillator

UJT Oscillator Waveforms

UJT-Controlled Relay Circuit

Junction Field Effect Transistor (JFET)

Field Effect Transistors (FET's) differ greatly from conventional transistors in that they are voltage-controlled, not current-controlled. Also, the input circuit is reverse-biased, therefore it offers a very high-input impedance, in some instances ranging to 50 thousand megohms.

A typical FET consists of a bar of lightly doped resistive n-type material having contacts at each end, with two etched areas of p-type material in the center. With an n-type bar the majority carriers are electrons. With the voltages shown in the illustration the electrons will enter the source connection and travel through the n-type material to leave at the drain connection. The source-to-drain current flow is "gated" or "channeled" between the two p-n junctions. Because of the junctions formed between the gate and channel the FET is often referred to as a "junction FET" or JFET. The JFET shown is an n-type, or n-channel JFET; JFET's using p-type material are referred to as p-type JFET's.

It was noted earlier in the book, in the discussion of reverse bias to the p-n junction of a diode, that increasing the value of reverse bias increases the value of the junction barrier. The junction barrier width increases because the increased reverse bias drives both holes and electrons further away from the junction. This barrier field is in effect an electrical field that drives away carriers, reducing current flow and increasing the resistance of the material in the barrier field. Application of a slight value reverse bias voltage to the gate, shown in A on the next page, will provide a small junction barrier about the p-n junction, with the junction barrier field projecting into the channel. With the "gating" action the area of the channel in which majority current carriers can flow has been reduced. The result is a corresponding reduction in source-to-drain current flow.

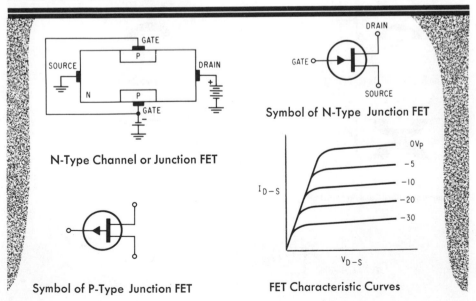

N-Type Channel or Junction FET

Symbol of N-Type Junction FET

Symbol of P-Type Junction FET

FET Characteristic Curves

Junction Field Effect Transistor (JFET) (Cont'd)

The action of the barrier field in depleting the number of available carriers in the field results in a <u>depletion region</u> (A) which is the common name for the barrier field.

As the value of reverse bias voltage at the gate circuit is increased, the area of the depletion region increases and further restricts source-to-drain current flow (B).   The consequence of gate bias voltage control is control over the source-to-drain current flow.   Increasing the gate reverse bias voltage to a point where the junction barrier fields overlap to "pinch off" and cause a complete cessation of channel current flow is illustrated in C.   The value of pinch-off voltage is labelled as V_P on the previous page. The voltage-current characteristic curves of an FET are seen to be similar to those of a vacuum tube.

Operation of a p-type JFET is identical to that of an n-type.   The polarities of the operating voltages are reversed and the majority current carriers through the p-type channel are holes.   Because the mobility of electron carriers is greater than that of hole carriers, n-type JFET's are more common than p-type because they provide higher efficiency at higher frequencies.

A typical JFET amplifier circuit is illustrated. As with a tube-type amplifier the input impedance is high and the load in the drain circuit is relatively high in value.   The voltage drop across the resistor in the source circuit is used as self-bias.   A circuit of this type would be used for the low-level high-impedance output of a device such as a crystal microphone.

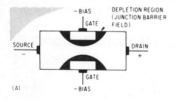

Depletion Region with Low Value Gate Bias

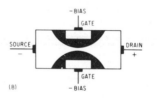

Medium Value Bias Increasing Depletion Region

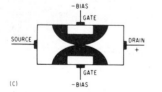

Maximum Value Bias Causing "Pinchoff"

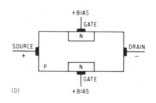

Construction of P-Type Junction FET

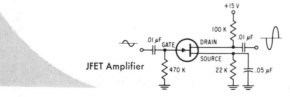

JFET Amplifier

Metal Oxide Semiconductor Field Effect Transistor (MOSFET)

MOSFET's are constructed using special processes similar to those used for processing integrated circuits, which are described later. Note in the illustration (and the symbol) that the gate circuit is insulated from the current carrying p-type material, thus the device is referred to as an Insulated Gate Field Effect Transistor (IGFET). Where the insulating material is metal oxide over silicon the transistor is referred to as Metal Oxide Semiconductor FET or MOSFET. Operation of both the IGFET and MOSFET are identical; it is only in their construction (the material used as an insulator) that they differ. Since MOSFET's are in greater use we will refer to this type in our discussion. In the symbol of a MOSFET the gate insulation is represented by a space between the gate and the bar representing the base (the base material is referred to as the "substrate"). The added insulation of the insulated gate provides MOSFET's with input resistances in the neighborhood of 500,000 megohms or higher, compared to the typical input resistance of a JFET of 50 thousand megohms.

With bias voltage applied to the gate circuit, the open gate circuit draws no current. However, the positive attraction of the bias source, as in a capacitor, pulls electrons away from the gate material. The gate is then positive (by the same value as the source) and is short electrons. The positively-charged gate will distort the orbit of the atoms of the metal oxide insulator, which is the dielectric. The positive charge of the gate attracts the electrons of the metal oxide dielectric atoms toward the gate and repels the positive nuclei of the atoms away from the gate and closer to the p-type substrate. The positive attraction of the metal oxide dielectric at the substrate attracts electrons (the minority carriers) in the p-type substrate to form (to "induce") a channel of minority carriers between the source and drain. The channel being of negative electrons can be referred to as an n-channel. Creation of a "channel" of carriers between the source and drain allows a current to flow between them.

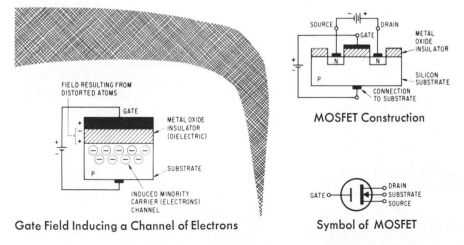

Gate Field Inducing a Channel of Electrons   MOSFET Construction

                                                        Symbol of MOSFET

Metal Oxide Semiconductor Field Transistor (MOSFET) (Cont'd)

The gate controls the source-to-drain current flow by controlling the number of minority carriers in the channel. Increasing the positive voltage on the gate will increase the number of attracted minority carriers in the channel; reducing channel resistance and increasing channel current flow. Decreasing the value of the gate's positive voltage will have the opposite effect.

Reversing the types of materials and polarities of applied voltage will provide a p-channel MOSFET. The minority carriers in the n-type substrate will be holes creating a p-channel of holes as current carriers between the source and drain.

Dual-gate JFET's and MOSFET's (see the symbols) are available for use in circuits where two signals are to be mixed, such as in converters.

To distinguish FET's from conventional transistors, FET's are often referred to as <u>unipolar</u> devices. Note that in both the JFET and MOSFET there is no gate current flow, the only current flow is from source to drain, thus the name "unipolar." In transistors we have minority carriers flowing in one direction and majority carriers flowing in another direction. For example, in an n-p-n transistor, electrons (the majority carriers) flow from emitter to collector; the holes (minority carriers) flow in the opposite direction, thus a transistor is referred to as a <u>bipolar</u> device.

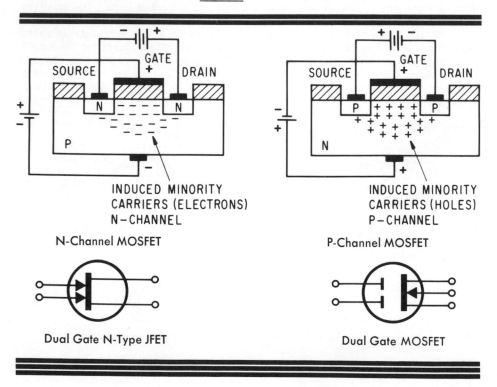

INDUCED MINORITY CARRIERS (ELECTRONS) N-CHANNEL

INDUCED MINORITY CARRIERS (HOLES) P-CHANNEL

N-Channel MOSFET

P-Channel MOSFET

Dual Gate N-Type JFET

Dual Gate MOSFET

Integrated Circuit Types

Integrated circuits are a natural result of the designers efforts to incorporate the construction of components along with the construction of the transistors and diodes.   By "integrating" all of the circuit components and placing them on one device, the manufacturing effort is substantially simplified, costs are reduced, power requirements are reduced, and a more reliable product is obtained. There are various methods of constructing integrated circuits. The most compact integrated circuit is the "monolithic" type in which all of the circuit components, including diodes and transistors, are integrated on one (mono) base, usually silicon. The base is most often referred to as a "substrate."

Monolithic integrated circuits are manufactured using highly complex methods. However, as shown in the illustration, the result is a tiny substrate (typical is 40 thousandths of an inch square, 6 thousandths of an inch thick), containing a complete complex circuit.

A simpler method of integrated circuit construction is by a "thick film" method in which the circuit is "printed" on a substrate approximately 1 inch by 1-1/4 inches, and 1/8 to 1/4 of an inch in thickness. Thick film circuits are constructed by printing the components such as resistors, capacitors, and inductors, then mounting the diodes and transistors on the substrate and wiring the diodes and transistors to the thick film printed circuit components.

Another method of construction is called, "thin film." As the name implies the thin film layers are only millionths of an inch in thickness. These thin layers are often deposited on monolithic circuits in the form of resistors and capacitors.

A "hybrid" method is to use a thick film integrated circuit with monolithic integrated circuits mounted on the thick film substrate and wired into the thick film printed circuit to form a complex but very compact "hybrid" integrated circuit.   Another form of hybrid is a monolithic circuit using thin film components.

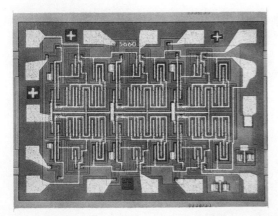

TYPICAL INTEGRATED CIRCUIT

(Courtesy RCA Corp.)

IC Linear Amplifier

The basic linear amplifier (also referred to as a "differential" amplifier) circuit shown is a d-c-coupled amplifier requiring two carefully matched transistors, matched base and collector resistors, and a single emitter resistor. Because of these matching requirements this amplifier circuit using discrete components is rarely used in communications circuits. Integrated circuit construction, however, overcomes these problems by placing resistors and transistors in a small area under carefully controlled conditions to provide inexpensive linear amplifiers using closely matched components.

The circuit can be used as an ordinary amplifier by applying the signal to one input (it doesn't matter which, A or B) and placing the other input to ground, or common. With an input signal to A, and input B at common, Q1 acts as a phase splitter, an inverted signal is available at output A; the in-phase emitter signal is directly coupled to the emitter of Q2. Transistor Q2 acts as a common-base amplifier; the in-phase amplified signal being available at the collector of Q2. The result is two out-of-phase signals at outputs A and B. Circuit operation is identical if the signal were to be applied at input B, with input A at common. The results would be identical with the output signals 180° out-of-phase with each other as shown in the illustration.

With signals applied to both inputs, the outputs will reflect the "difference" between the inputs, with the outputs being 180° out-of-phase with each other. For example, should input A have a +0.3V signal and input B a +0.1V signal, the outputs will reflect the difference; it would be as though a +0.2V signal were applied. Used in this manner it is referred to as a differential amplifier. We see then that the amplifier can be used in four ways: 1 input, 1 output; 1 input, 2 outputs; 2 inputs, 1 output; 2 inputs, 2 outputs; with the outputs always being 180° out-of-phase.

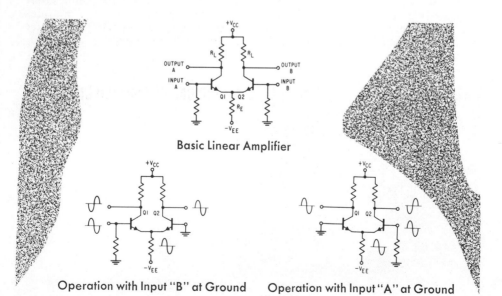

Basic Linear Amplifier

Operation with Input "B" at Ground   **Operation with Input "A" at Ground**

A Multi-Purpose Linear IC Amplifier

The RCA type 3020 IC is a multi-purpose linear integrated circuit that can be used in audio circuits as a preamplifier, phase inverter, driver, or low-power output stage. It operates from a 9-volt supply, and is often used in miniature portable equipment operating from a 9-volt "transistor" battery.

The circuit contains an extra transistor, Q1, that can be used as an amplifier or for isolation (buffer) by applying the input signal to pin 10 and with an emitter resistor from pin 1. The output of Q1 (used as an emitter-follower) can be applied to either, or both, inputs (pins 2 and 3). The circuit contains a series of diodes that act as a voltage divider and, at the same time, they provide thermal compensation to prevent gain fluctuations due to temperature variations.

Transistors Q2 and Q3 form the linear amplifier. The collector output of Q2 is direct-coupled to the base of Q4, and the collector output of Q3 is direct-coupled to the base of Q5, with Q4 and Q5 acting as drivers to the output circuit of Q6 and Q7. The emitter output of Q4 is fed back via R5 to the base of Q2, and the emitter output of Q5 is fed back via R7 to the base of Q3 to provide negative feedback. This feedback improves the frequency response and helps to maintain a constant gain level within the integrated amplifier circuit. Transistors Q6 and Q7 are output power transistors that drive the external circuit.

The circuit illustrated shows the RCA 3020 used to drive a low-impedance speaker in a complete audio amplifier circuit.

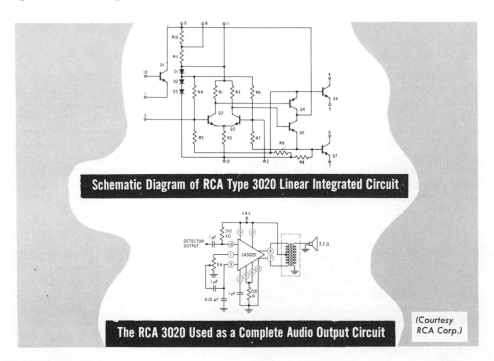

Schematic Diagram of RCA Type 3020 Linear Integrated Circuit

The RCA 3020 Used as a Complete Audio Output Circuit

(Courtesy RCA Corp.)

Construction

An example of the methods used in creating a monolithic integrated circuit is illustrated by the various steps and processes that form a single n-p-n transistor. It must be noted that during the processes more than just the one transistor is being formed; diodes may be formed as well as resistors, capacitors, or inductors. In the illustration, step (a) represents a portion of a substrate wafer of n-type silicon. A layer of silicon dioxide is then formed over the wafer, step (b). Next, a drop of light-sensitive fluid, called photo-resist, is placed on top of the wafer. The wafer is then spun at high speed to have the photo-resist form a thin layer on top of the silicon-dioxide, step (c). A "mask" is then placed over the substrate to expose certain portions of the substrate to ultra-violet light and "mask" or prevent other portions of the substrate from being affected by the ultra-violet light, step (d). The mask is then removed, step (e). Next, the substrate is placed in a developer which hardens those areas of the photo-resist exposed to the ultra-violet light while removing the masked, unexposed portions of the photo-resist, step (f).

The remaining photo-resist forms a pattern, over which a chemical is placed to "etch" the exposed silicon dioxide, in turn leaving a portion of the substrate exposed, step (g). Then the photo-resist is removed by special chemicals, step (h), and the substrate placed in a diffusion oven. After leaving the diffusion oven we find the exposed portion of the substrate has had p-type impurities diffused into it to form a p-n junction, step (i). At this point in the process a diode has been formed.

Continuing the process, the silicon dioxide substrate covering is removed by special chemicals, step (j), and a new layer of silicon is placed over the substrate, step (k). The process is repeated, using a new mask to expose different areas, step (l). The mask is removed and the substrate "developed" to expose different areas of the silicon dioxide. The silicon dioxide is then etched to expose the diffused p-area of the substrate, step (m). The diffusion process is then repeated, the photo-resist is removed and the substrate is placed in a diffusion oven, but this time the impurity is n-type to form a diffused area of n-type material in the center of the p-type area. The result is an n-p-n transistor, step (n).

The final process is to again strip the substrate clean of silicon dioxide, and to place a new layer of silicon dioxide and photo-resist on the substrate; a new mask is used and the substrate exposed, step (o). The previously explained steps are repeated to leave the etched layer of silicon dioxide as shown in step (p). This is then covered by a layer of metal to interconnect the various diodes, transistors, and components, step (q). The excess metal is then etched and removed by a masking process (r). The remaining metal interconnections form the desired integrated circuit.

INTEGRATED CIRCUITS

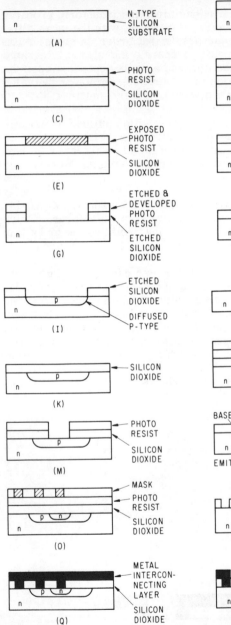

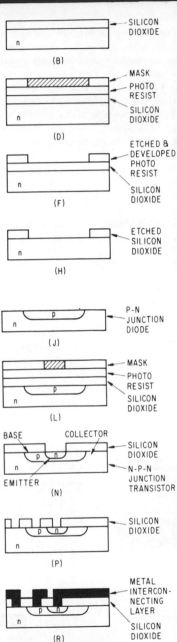

Components

Of the three basic components, resistors, capacitors, and inductors, probably resistors are the simplest to form in integrated circuits, with the more difficult components being capacitors and inductors, in that order. A resistor can be diffused during one of the steps required to prepare a monolithic integrated circuit. The value of resistance of the diffused area is dependent upon the doping level (the greater the concentration of impurities, the lower the value of resistance), the cross section (depth and width), and the length.

Another method of placing a resistor on a substrate, either silicon or ceramic, is by use of a thin film. In the thin film process the entire substrate is coated with a thin film, or layer, of high resistance metal. This layer is then masked and the undesired portions removed by chemical "etching" to leave the thin film resistors in the desired areas.

For monolithic integrated circuits constructed on a silicon substrate, resistors can be processed by diffusion, epitaxial growth, or thin films. Thick film integrated circuits can be processed by screening resistive ink or by thin film plating to create resistors on the substrate.

Forming a p-n junction (of the type illustrated) forms a capacitor. The junction barrier is effectively a dielectric separating the n- and p-type materials, forming a capacitor. It was noted in the discussion of junction diode reverse bias that increasing the value of reverse bias increased the junction barrier. Consequently we can increase or decrease the width of the barrier junction (the dielectric) between the n- and p-type materials of the junction by increasing or decreasing the value of reverse bias. This points out that a junction capacitor must have reverse bias to be effective and that the value of reverse bias must be held steady to prevent undesired changes in capacity.

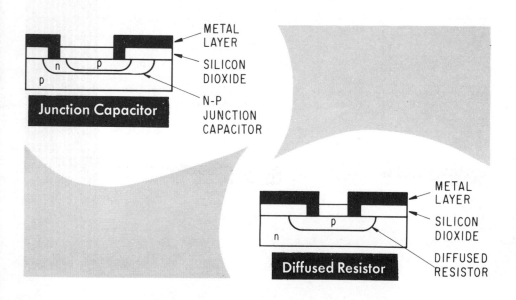

Zener diodes break down and conduct heavily when the correct value of reverse voltage is applied to the diode.

Tunnel diodes exhibit a "negative conductance" region which allows amplification of very low level signals.

Silicon controlled rectifiers pass no current until gated to turn on, they then conduct large values of current.

The fixed value of voltage required to have a unijunction transistor conduct permits its use as a "trigger" to control other circuits.

Field Effect Transistors (FET's) differ greatly in that they offer very high input resistance.

Because the mobility of electron carriers is greater than that of hole carriers n-type FET's are in more common use than p-types.

Metal Oxide Semiconductor Field Effect Transistors (MOSFET's) have an isolated gate to provide unusually high values of input resistance.

Field Effect Transistors are referred to as "unipolar" devices, conventional junction transistors as "bipolar" devices.

Integrated circuits of the "monolithic" type have all components constructed on one substrate at the same time to develop a complete circuit measuring as little as 40 thousandths of an inch square by six thousandths of an inch high.

Thick film circuits are easier to construct using special "printing" processes to develop a circuit containing all components less diodes and transistors on a 1 inch by 1 1/4 inch substrate, approximately 1/4 of an inch thick.

Hybrid circuits are formed by combining monolithic integrated circuits with thick film or thin film integrated circuits.

Linear (differential) amplifier circuits can be economically constructed using integrated circuit techniques.

Differential amplifier circuits provide a wide combination of input and output arrangements.

The use of reverse biased diode junctions as capacitors requires that the values of bias voltage be held constant to prevent undesired changes in capacity.

REVIEW QUESTIONS

1. Explain how a zener diode can be used to regulate voltage.
2. In a tunnel diode, what is meant by "negative conductance"?
3. In an SCR, what is meant by "forward breakover voltage"?
4. In a unijunction transistor how is the B1-B2 current controlled?
5. Describe the basic reason for FET's offering such high input resistance.
6. Explain the operation of the gate bias voltage in controlling FET current flow.
7. What is the difference between unipolar and bipolar transistors?
8. How thin are the layers of a thin film circuit?
9. Explain how a differential amplifier can provide the different input and output combinations.
10. Briefly, describe how monolithic integrated circuits are constructed.
11. In a monolithic integrated circuit, what determines the value of a diffused resistor?

GLOSSARY

Acceptor: A substance with three electrons in the outer orbit of its atom which when added to a semiconductor crystal, provides a hole in the lattice structure of the crystal. An acceptor (indium or gallium) is a p-type impurity.

Alpha (α): The current gain factor of a transistor when connected in a common-base circuit. Alpha is equal to the ratio of collector current change to emitter current change for a constant collector voltage.

Alpha Cutoff Frequency: The frequency at which the alpha of a transistor falls below 0.707 times the maximum gain.

Barrier: The electric field between the acceptor ions and the donor ions at a junction.

Barrier Height: The difference in potential from one side of a barrier to the other.

Base: The center semiconductor material of a double junction (n-p-n or p-n-p) transistor. The base is comparable to the grid of an electron tube.

Beta (β): The current gain factor of a transistor connected in a common-emitter circuit. Beta is equal to the ratio of a change in collector current to a change in base current for a constant collector voltage.

Bias: The d-c operating voltage or current applied to an element of a transistor. Bias current establishes the operating point of a transistor.

Collector: The end semiconductor material of a double junction (n-p-n or p-n-p) transistor that is normally reverse-biased with respect to the base. The collector is comparable to the plate of an electron tube.

Common-Base Amplifier: A transistor amplifier in which the base is common to the input and output circuits. This circuit is comparable to the grounded-grid triode circuit.

Common-Collector Amplifier: A transistor amplifier in which the collector is common to the input and output circuits. This circuit is comparable to the cathode follower electron-tube circuit.

Common-Emitter Amplifier: A transistor amplifier in which the emitter is common to the input and output circuits. This circuit is comparable to the conventional common-cathode electron-tube circuit.

Complementary Symmetry: An arrangement of p-n-p and n-p-n transistors that provides push-pull operation from a-single input signal. Such a circuit makes use of the similar, but opposite characteristics of p-n-p and n-p-n transistors.

Diode: A p-n junction which when reverse-biased exhibits high resistance and when forward-biased exhibits low resistance.

Direct Coupling: Coupling the output of an amplifier stage directly to the input of the succeeding stage, eliminating coupling circuit frequency losses.

Donor: A substance with electrons in the outer orbit of its atom which, when added to a semiconductor crystal, provides a free electron in the lattice structure of a crystal. A donor is an n-type impurity. Typical donors are antimony and arsenic.

Electron-Pair Bond: A valence (covalent) bond formed by two electrons, one from each of two adjacent atoms.

Emitter: The end semiconductor material of a double junction (p-n-p or n-p-n) transistor that is forward-biased with respect to the base. The emitter is comparable to the cathode of an electron tube.

Forward Bias: An external potential applied to a p-n junction so that the barrier is lowered and relatively high current flows through the junction.

Heat Sink: A mass of metal or other good heat conductor used to rapidly dissipate the heat energy produced by a transistor.

Hole: A mobile vacancy in the electonic valence structure of a semiconductor. The hole acts similarly to a positive electronic charge having a positive mass.

Hybrid Integrated Circuit: Integrated circuits formed by combining monolithic integrated circuits with thick or thin film integrated circuits.

Impurity: A substance added to a semiconductor to give it a p-type or n-type characteristic.

Integrated Circuits: Specially constructed circuits utilizing the same methods of construction for transistors, diodes, capacitors, resistors, and inductors enabling the complete circuit to be placed on one base.

Junction: A point or area of contact between n- and p-type semiconductors.

Junction Field Effect Transistor (JFET): A voltage-controlled transistor providing very high input resistances.

Junction Transistor: A device having three alternate sections of p-type or n-type semiconductor material.

Lattice Structure: In a crystal, a stable arrangement of atoms and their electron-pair bonds.

Majority Carriers: The holes in p-type semiconductors or free electrons in n-type semiconductors.

Metal Oxide Semiconductor Field Effect Transistor (MOSFET): A field effect transistor using a special insulated gate circuit to provide exceptionally high input resistance.

Minority Carriers: The holes in n-type semiconductors or excess electrons in p-type semiconductors.

Monolithic Integrated Circuit: An integrated circuit in which all circuit components are simultaneously constructed on one base.

N-P-N Transistor: A device consisting of a p-type section and two n-type sections of semiconductor material, with the p-type in the center.

N-Type Semiconductor: A semiconductor crystal into which a donor impurity has been introduced. It contains free electrons.

P-N Junction: The area of contact between n-type and p-type semiconductor materials.

P-N-P Transistor: A device consisting of an n-type section and two p-type sections of semiconductor material, with the n-type in the center.

Point Contact: A physical connection made by a metallic wire on the surface of a semiconductor.

P-Type Semiconductor: A semiconductor crystal into which an acceptor inpurity has been introduced. It provides holes in the crystal lattice structure.

Reverse Bias: An external potential applied to a p-n junction to raise the barrier and prevent the movement of majority current carriers.

Saturation (Cutoff) Current: The current flow between the base and collector or between the emitter and collector, measured with the emitter lead or the base lead open.

Semiconductor: A conductor whose resistivity is between that of metals and insulators. It exists in crystalline form.

Silicon Controlled Rectifier (SRC): A four-layer diode with a special gate lead. The SRC is essentially a controlled rectifier that will not pass current until "turned-on" by a gate signal.

Stabilization: The reduction of variations in voltage or current due to undesirable circuit changes.

Thick Film Integrated Circuits: Using special "printing" processes develops an integrated circuit that by comparison is thicker than the thin film or monolithic type integrated circuits.

Thin Film Integrated Circuits: Thin films, usually millionths of an inch thick, are used to form integrated circuit components.

Transistor: A semiconductor device capable of transferring a signal from one circuit to another and producing amplification.

Tunnel Diode: A special diode named after the apparent "tunneling" effect that takes place at the junction to provide a negative conductance region.

Unijunction Transistor (UJT): A special three-terminal device containing one junction that will conduct at a "triggering" level to have the UJT turn-on and allow heavy current flow.

Zener Diode: A p-n junction diode reverse-biased into the breakdown region, used for voltage stabilization.

Index

INDEX
vol. 5

(Note: A cumulative index covering all six volumes in this series is included at the end of Volume 6.)